Novel Nanomaterials for Thermochemical Storage

Novel Nanomaterials for Thermochemical Storage: Development and Characterization

Editor

Candida Milone

MDPI • Basel • Beijing • Wuhan • Barcelona • Belgrade • Manchester • Tokyo • Cluj • Tianjin

Editor
Candida Milone
Universita degli Studi di Messina
Italy

Editorial Office
MDPI
St. Alban-Anlage 66
4052 Basel, Switzerland

This is a reprint of articles from the Special Issue published online in the open access journal *Nanomaterials* (ISSN 2079-4991) (available at: https://www.mdpi.com/journal/nanomaterials/special_issues/nano_thermo_storage).

For citation purposes, cite each article independently as indicated on the article page online and as indicated below:

LastName, A.A.; LastName, B.B.; LastName, C.C. Article Title. *Journal Name* **Year**, *Volume Number*, Page Range.

ISBN 978-3-0365-0758-3 (Hbk)
ISBN 978-3-0365-0759-0 (PDF)

© 2021 by the authors. Articles in this book are Open Access and distributed under the Creative Commons Attribution (CC BY) license, which allows users to download, copy and build upon published articles, as long as the author and publisher are properly credited, which ensures maximum dissemination and a wider impact of our publications.

The book as a whole is distributed by MDPI under the terms and conditions of the Creative Commons license CC BY-NC-ND.

Contents

About the Editor vii

Preface to "Novel Nanomaterials for Thermochemical Storage: Development and Characterization" ix

Salvatore Vasta, Vincenza Brancato, Davide La Rosa, Valeria Palomba, Giovanni Restuccia, Alessio Sapienza and Andrea Frazzica
Adsorption Heat Storage: State-of-the-Art and Future Perspectives
Reprinted from: *Nanomaterials* **2018**, *8*, 522, doi:10.3390/nano8070522 1

Alenka Ristić and Nataša Zabukovec Logar
New Composite Water Sorbents $CaCl_2$-PHTS for Low-Temperature Sorption Heat Storage: Determination of Structural Properties
Reprinted from: *Nanomaterials* **2019**, *9*, 27, doi:10.3390/nano9010027 15

Sergio Salviati, Federico Carosio, Guido Saracco and Alberto Fina
Hydrated Salt/Graphite/Polyelectrolyte Organic-Inorganic Hybrids for Efficient Thermochemical Storage
Reprinted from: *Nanomaterials* **2019**, *9*, 420, doi:10.3390/nano9030420 31

Georg Gravogl, Christian Knoll, Jan M. Welch, Werner Artner, Norbert Freiberger, Roland Nilica, Elisabeth Eitenberger, Gernot Friedbacher, Michael Harasek, Andreas Werner, Klaudia Hradil, Herwig Peterlik, Peter Weinberger, Danny Müller and Ronald Miletich
Cycle Stability and Hydration Behavior of Magnesium Oxide and Its Dependence on the Precursor-Related Particle Morphology
Reprinted from: *Nanomaterials* **2018**, *8*, 795, doi:10.3390/nano8100795 43

Elpida Piperopoulos, Marianna Fazio and Emanuela Mastronardo
Synthesis of Me Doped $Mg(OH)_2$ Materials for Thermochemical Heat Storage
Reprinted from: *Nanomaterials* **2018**, *8*, 573, doi:10.3390/nano8080573 57

Stefania Doppiu, Jean-Luc Dauvergne and Elena Palomo del Barrio
Solid-State Reactions for the Storage of Thermal Energy
Reprinted from: *Nanomaterials* **2019**, *9*, 226, doi:10.3390/nano9020226 75

About the Editor

Candida Milone (Prof.) is Full Professor of Chemistry at the University of Messina (Italy). She completed her PhD in "Chemistry of Materials" in 1993. She has been Head of the Department of Engineering at the University of Messina since 2018. She has been a member of the European Panel of the International Energy Agency (IEA) -Task 58 "Material and Component Development for Thermal Energy Storage" since 2017. Her scientific activity concerns the design, synthesis and characterization of nanostructured materials for catalysis, environmental and energy applications. Currently, she is working on the development of more efficient and durable Thermochemical Energy Storage Materials for low- and middle-temperature waste heat storage, produced by renewable energy production plants.

Preface to "Novel Nanomaterials for Thermochemical Storage: Development and Characterization"

The aim of this book is to present the latest advances in the development of ThermoChemical heat storage Materials (TCM). TCM, actually, represent a key factor for boosting the use of renewable energy in heating and cooling systems effectively, as they decouple the availability of renewable energy from the time when it is needed. Moreover, the use of this emerging technology contributes to improve the energy efficiency of global energy systems by storing waste heat. The development of thermochemical materials and technologies is still at an early stage. It is imminently necessary to improve some fundamental aspects of the use and commercialization of these systems, such as power, energy density, efficiency and cost. Furthermore, environmental friendliness must be considered. This book, deepening the characteristics of known or innovative TCM, aims to be an aid in the comparison of the most promising and best-performing materials for advancement in their application fields. The first work "Adsorption Heat Storage: State-of-the-Art and Future Perspectives" reports a literature review of the recent advancements in the field of adsorption Thermal Energy Storage (TES) systems. The collected articles refer to the full range of applicability of thermochemical energy storage materials, low-, medium- and high-temperature. The article "New Composite water Sorbents $CaCl_2$-PHTS for Low-Temperature Sorption Heat Storage: Determination of Structural Properties" focuses attention on sorbents operating in the range between 80 and 150 °C; in particular, it investigates a plugged hexagonal templated silicate matrix filled with $CaCl_2$. "Hydrated Salt/Graphite/Polyelectrolyte Organic-Inorganic Hybrids for Efficient Thermochemical Storage" deals with the preparation and characterization of strontium bromide hexahydrate/graphite composite material for seasonal storage applications. For medium-temperature applications "Cycle Stability and Hydration Behavior of Magnesium Oxide and Its Dependence on the Precursor-Related Particle Morphology" shows the correlation between the structural characteristics of MgO and cycle stability. Furthermore, metal doping during $Mg(OH)_2$ synthesis is investigated in "Synthesis of Me Doped $Mg(OH)_2$ Materials for Thermochemical Heat Storage" to increase volumetric stored/released heat capacity. The article "Solid-State Reactions for the Storage of Thermal Energy" reports the use of solid-state reactions for the storing of thermal energy at high temperatures; in particular, the candidate reactions, eutectoid and peritectoid type transitions, are considered. The present book is meant to be a valuable support for researchers operating in the field of innovative thermochemical materials, to provide a general view of the current development stage on innovative TCM at low, medium and high temperatures and to offer a discussion on probable and practicable solutions for the feasible use of such systems.

Candida Milone
Editor

Review

Adsorption Heat Storage: State-of-the-Art and Future Perspectives

Salvatore Vasta *, Vincenza Brancato, Davide La Rosa, Valeria Palomba, Giovanni Restuccia, Alessio Sapienza and Andrea Frazzica

CNR—ITAE, Istituto di Tecnologie Avanzate per l'Energia "Nicola Giordano", Via Salita S. Lucia sopra Contesse 5, 98126 Messina, Italy; vincenza.brancato@itae.cnr.it (V.B.); davide.larosa@itae.cnr.it (D.L.R.); valeria.palomba@itae.cnr.it (V.P.); giovanni.restuccia@itae.cnr.it (G.R.); alessio.sapienza@itae.cnr.it (A.S.); andrea.frazzica@itae.cnr.it (A.F.)
* Correspondence: salvatore.vasta@itae.cnr.it; Tel.: +39-090-624404

Received: 24 June 2018; Accepted: 7 July 2018; Published: 12 July 2018

Abstract: Thermal energy storage (TES) is a key technology to enhance the efficiency of energy systems as well as to increase the share of renewable energies. In this context, the present paper reports a literature review of the recent advancement in the field of adsorption TES systems. After an initial introduction concerning different heat storage technologies, the working principle of the adsorption TES is explained and compared to other technologies. Subsequently, promising features and critical issues at a material, component and system level are deeply analyzed and the ongoing activities to make this technology ready for marketing are introduced.

Keywords: adsorption; heat storage; thermo-chemical; zeolite; silica gel; adsorbent materials

1. Introduction

One of the key technologies for boosting the diffusion of renewable energies and for developing efficient energy systems is Thermal Energy Storage (TES). Indeed, the employment of TES has the potential to overcome the existing mismatch between energy production and demand for discontinuous energy sources (e.g., solar thermal) and/or variable loads (e.g., thermal energy demand in buildings). It is therefore evident that this component is gaining a crucial role in the development of highly efficient thermal energy systems [1]. The three main technologies for thermal energy storage are: sensible, latent and thermo-chemical heat storage, as shown in Figure 1.

In a sensible heat storage, thermal energy is stored as a function of the temperature difference only. The amount of stored energy depends both on the specific heat and on the temperature difference between the charge and discharge phase. The heat storage media can be either liquid or solid. The most common example of sensible heat storage media is water. Indeed, it represents the typical medium employed for heat storage at temperatures below 100 °C, as it couples its abundance, cost-effectiveness, and non-polluting features with good thermodynamic characteristics, such as high specific heat and good heat transfer efficiency within the operating ranges. However, its working temperature range is limited between 0 and 100 °C, and it presents corrosive behavior.

The main advantages of sensible heat storages are related to the simple design, cost-effectiveness, and wide temperature range of applications. Nevertheless, they are characterized by low heat storage density and degradation of the stored energy due to heat dissipation through the environment. This means that they cannot be efficiently used for storing heat for medium/long-term periods [2].

In the latent heat technology storage, thermal energy is mainly stored by exploiting the latent heat of the phase transition of the heat storage medium. The phase change process can involve either a solid/liquid transition (i.e., melting/solidification process) or a solid/solid transition (i.e., transition from one crystal structure to a different one). The most common class of materials employed as latent heat storage media are: salts, water, hydrated salts and paraffins, which usually undergo a solid/liquid transition. The advantages of such a technology are higher heat storage density compared to sensible systems, the possibility to store energy in a narrow temperature range, and the possibility to employ the phase transition to smooth temperature fluctuations (e.g., in building envelope applications). The main disadvantages are related to the slow kinetics of the phase transition, which limits the charging/discharging power, the instability of materials undergoing several melting/solidification processes, and the presence of a volume variation when passing from the solid to liquid phase, which needs to be carefully considered during the design process. Furthermore, latent heat storage suffers from heat dissipation to the environment, which limits its application for long-term heat storage [3].

The thermo-chemical technology is based on the reversible reaction occurring between two components and is associated with a high amount of energy. These reactions can be either chemical or physical. The main limits are related to the very slow reaction kinetics, due to the high energy associated with the process, as well as heat and mass transfer diffusion resistance within the material. Physical reactions are typical of sorption applications, where a refrigerant (e.g., water, ammonia) reacts with a sorbent, which can be either liquid (absorption systems), or solid (adsorption systems). As this technology is based on physical reaction, it generally needs lower charging temperatures (i.e., 70–150 °C) and is characterized by lower reaction enthalpies compared to the chemical reactions. Accordingly, they are characterized by faster kinetics but lower heat storage densities [4].

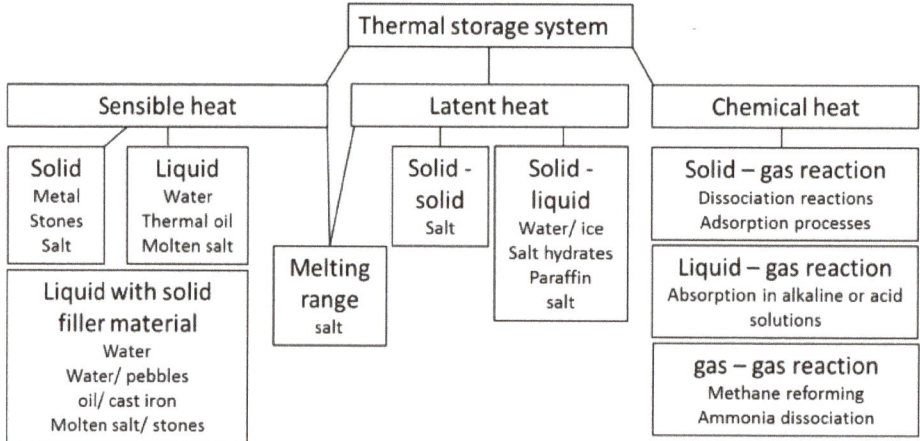

Figure 1. Thermal energy storage (TES) technologies [5].

The present paper will focus on the analysis of the state-of-the-art and future perspectives of the adsorption heat storage systems, mainly for domestic and tertiary sectors.

2. Principles of Adsorption Heat Storage

Adsorption heat storage belongs to the wider class of thermo-chemical heat storage. This technology is based on the interaction between a liquid sorbate, usually water, and a solid sorbent (e.g., zeolites, silica gels, activated carbons). This interaction occurs between the sorbate molecules and the available surface of the solid, as represented in Figure 2.

In order to explain the working principle of an adsorption heat storage, it is necessary to distinguish between direct and indirect heat storage methodologies [6]. The direct heat storage, as represented in Figure 3, is the typical technology employed to store heat in both sensible and latent form.

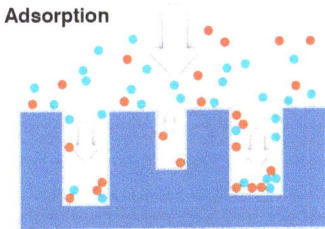

Figure 2. Adsorption of refrigerant over the external surface of an adsorbent solid material.

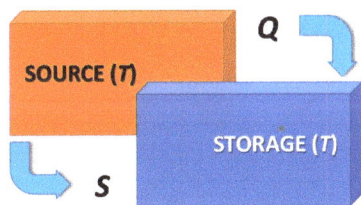

Figure 3. Direct heat storage, charging phase.

In this case, there is a coupled heat and entropy flux from the heat source to the storage, at temperature T. Accordingly, since heat and entropy are closely related, Q = T·S, this means that the heat storage capacity depends on the temperature level and on the entropy content. Due to the temperature difference between TES (either hot or cold storage) and ambient temperature, there is a continuous loss of heat-storage capacity over time, due to the reduction of ΔT between stored heat and ambient temperature. Accordingly, it is possible that below certain levels of ΔT, the stored heat is no longer useful for practical applications (e.g., domestic hot water, space heating).

In contrast, an indirect TES, as represented in Figure 4 for the charging process, can overcome this limitation, converting heat into a different form of energy (e.g., mechanical, chemical), which is stored without any limitation (i.e., no heat losses to the ambient). As this technology is based on an energy conversion process, the converter needs to be connected to an external sink (e.g., the ambient) through which the produced waste heat and entropy, due to irreversibility, must be dissipated. The reverse phenomena, which exploits heat and entropy flux from the ambient to convert the stored energy into useful heat again, represents the discharging phase. It is therefore evident that this technology needs to be connected to two different sources/sinks, which makes the TES an indirect process, like a heat pump.

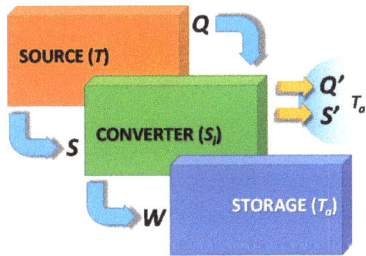

Figure 4. Indirect heat storage, charging phase.

Adsorption heat storages belong to the indirect TES class. Indeed, in this case, heat is employed to drive a desorption process, which means that energy is stored in the form of adsorption potential energy. In this way, heat is stored and conserved until the refrigerant fluid (adsorbate) is kept separate from the adsorbent.

Generally, there are two system configurations for adsorption TES: closed- and open-cycle.

Figure 5 reports the working phases of a closed adsorption TES. During the charging phase as shown in Figure 5a, the adsorber, in which the adsorbent material is saturated with adsorbate, is regenerated exploiting heat, Q_{des}, coming from the heat source. The desorbed vapor condenses in the condenser, and the heat of condensation, Q_{cond}, is dissipated into the ambient or delivered to the load, if the temperature level is adequate. Once the charging process is completed and the adsorbent material is dry, the connection between condenser and adsorber is closed. In this condition, the system can conserve the stored energy for an indefinite time, since the thermal energy is stored as the adsorption potential between adsorbate and adsorbent material. To recover the stored thermal energy, as shown in Figure 5b, the connection between the liquid adsorbate reservoir, which in this phase acts as an evaporator, and adsorber is opened once again. During this discharging phase, the adsorbate is evaporated, adsorbing heat from the ambient, Q_{evap}, before the vapor flows into the adsorber, where it is adsorbed. Since the adsorption process is exothermic, the heat of adsorption, Q_{ads}, is released to the load.

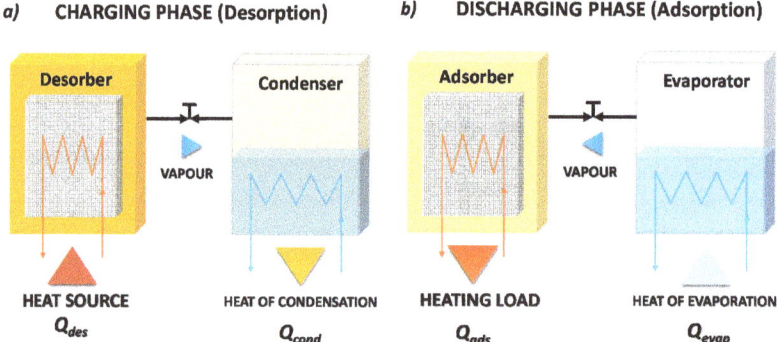

Figure 5. A closed adsorption heat storage cycle: (**a**) charging phase; (**b**) discharging phase.

Clearly, this process is defined as closed since the adsorbate is continuously condensed/evaporated in a closed system without any mass exchange with the ambient.

Contrastingly, the open adsorption TES system, represented in Figure 6, continuously exchanges mass (adsorbate) with the ambient. The two charging/discharging phases are, in fact, similar to those already described for the closed cycle. In this case, heat is provided and extracted by fluxing air through the adsorbent bed. Specifically, during the charging/desorption phase, a hot and dry air flux enters the storage, causing the desorption of adsorbed water, and exits at a lower temperature and higher humidity content.

During the discharging/adsorption phase, a humid and cooled air flux is provided to the dry adsorbent, which triggers the adsorption and consequent release of the stored heat. The stored heat is released as hot and dry air flux exiting the system. Some clear differences must be highlighted between closed and open adsorption TES.

- Open adsorption TES can employ only water as adsorbate, since they exploit the moisture of the ambient air as working fluid. On the contrary, closed adsorption TES can use a different adsorbate, even if the most widely employed is water, owing to its high latent heat as well as its environmental friendly feature.

- Open adsorption TES strongly depends on the external ambient conditions. This means that the higher the moisture content of the external air, the higher the heat storage density that can be achieved.
- Closed adsorption TES, if employed for short-term heat storage, can also exploit the heat-pumping effect, related to the energy recovered from the condensation of water vapor during the charging phase. In this manner, the energy storage density is enhanced. On the contrary, for the open adsorption TES, the heat of condensation is dumped to the ambient and not recovered.
- Closed adsorption TES are usually more complex systems, since they employ different heat exchangers to provide/extract heat to the adsorber and the evaporator/condenser. Furthermore, working in a closed cycle, they need to keep a saturated adsorbate atmosphere, meaning that any air leakage must be prevented, making systems more complex and expensive. On the contrary, open adsorption TES are less complex and expensive systems, which seem more suitable for long-term heat storage.

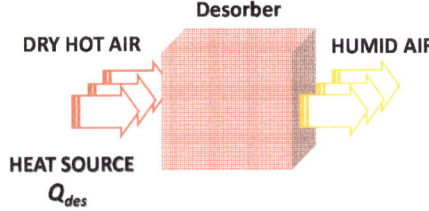

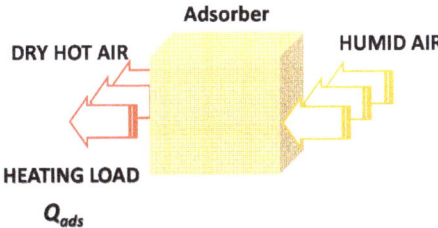

Figure 6. An open adsorption heat storage cycle: (a) charging phase; (b) discharging phase.

3. State-of-the-Art

Adsorption TES is considered to be quite a promising technology both for seasonal and daily storage applications, nevertheless, its commercial diffusion is still not fully developed, mainly due to its cost and the lack of technical knowledge at a system level. This means that there is still need for development and research, in order to make the technology commercially competitive. The research activities in the field can be divided into three levels: Materials, components, and systems.

3.1. Adsorbent Materials

Development of adsorbent materials for adsorption TES is strongly related to the adsorbate to be employed. As the most common adsorbates are water and ammonia, the following sections summarize the research activities on adsorbents developed for these kinds of adsorbates.

3.1.1. Silica Gels

Silica gels historically represent one of the most employed adsorbent materials for water vapor adsorption. In fact, they represent a less expensive option for adsorption TES applications and can be easily employed for heat sources at temperatures lower than 100 °C (e.g., flat-plate solar thermal collectors). It is important to highlight that the porous structure of silica gels for closed adsorption TES must be completely different from that which is employed for open adsorption TES. Indeed, as in a closed system the adsorption/desorption process usually occurs in a limited partial pressure range (e.g., between 0.1 and 0.3 p/p_0), it is necessary to have silica gels with highly microporous structures, capable of exchanging high quantities of water vapor. On the contrary, in an open adsorption TES, since the working partial pressures are usually higher, a mesoporous silica gel can be also employed, due to the capillary condensation phenomena that occurs within this working range.

As will be described later in this work, despite their competitive cost and wide availability, silica gels showed too low a heat storage capacity, which often leads to an experimental heat storage density even lower than that of water [7,8]. Nevertheless, they still represent a possible option if employed for long-term heat storage applications. With regard to such applications, the cost can become the main selection criteria. Furthermore, their application as a host matrix for composites is considered an interesting alternative.

3.1.2. Classical Zeolites

Zeolites are crystalline alumino-silicates, characterized by a high specific surface area (i.e., about 800 m^2/g) and wide microporous volumes, which make these materials perfectly suitable for water vapor adsorption. Owing to their porous structure, zeolites are usually highly hydrophilic, which allows them to obtain high adsorption capacities even at low partial pressures. This high affinity with water, of course, is reflective of strong bonding that requires higher temperatures to be broken compared to silica gels (i.e., more than 150 °C). Zeolites type A, 13X and Y are the most common classical synthetic zeolites employed for adsorption heat storage. These materials are mostly used for open adsorption TES, since, in order to get enough energy storage density, they must be regenerated at high temperatures, making air the most effective heat transfer medium. Thus, due to the required high operation temperature, they are usually employed for industrial waste heat recovery and storage [9].

Owing to their crystalline structure, as opposed to the amorphous structure of silica gels, they can guarantee a higher long-term hydrothermal stability; offering a more reliable option for applications where several adsorption and desorption cycles are expected.

3.1.3. Zeo-Like Materials

More recently, several new microporous adsorbents have been proposed for TES applications. They are often referred to as zeo-like materials, as their crystalline structure is somewhat similar to those of classical zeolites. The two classes that showed the most promising features are the aluminophosphates (AlPOs) and the silico-aluminophosphates (SAPOs). Indeed, in contrast to other classical adsorbents, these materials show a partially hydrophobic behavior, that is reflected in an S-shaped adsorption isotherm. This is an advantageous characteristic, that allows a high amount of water vapor exchange to be obtained in a narrow range of partial pressure. Accordingly, since the overall heat storage capacity is highly dependent on the water vapor exchange, these materials can guarantee very high heat storage capacities.

Among these two classes, the most attractive materials are known as AlPO-18 and SAPO-34, as reported by several authors [10]. Particularly, the research on these materials lead to the first commercial adsorbent specifically developed for closed adsorption systems (e.g., for heating, cooling and storage applications). It is known as AQSOA Z02, and is produced and commercialized by Mitsubishi Plastic Inc. (Chiyoda-ku, Tokyo) [11].

3.1.4. Metal Organic Frameworks

Metal Organic-Frameworks (MOFs) represent a new, emerging class of adsorbent materials [12]. These materials, still in an early stage of development, are considered the future of adsorption TES. Due to their structure, made up of metal ions interconnected by organic macro-molecules, it is possible to select several different compositions; giving infinite possibilities to obtain the ideal adsorbent material by simply adjusting the synthesis procedure. Indeed, they are usually characterized by a high specific surface area (i.e., higher than 2000 m^2/g), which guarantees the ability to reach higher adsorption capacities compared to other adsorbent classes. Higher adsorption capacities are also achieved by tuning of the pore sizes, according to the adsorbate and the working range. Nevertheless, as previously mentioned, this class is still far from practical application, due to two main reasons: Their high cost, related both to the small amount currently produced and to the cost of raw materials, and their hydrothermal stability, that still requires thorough investigation.

3.1.5. Activated Carbons

Activated carbons are carbonaceous adsorbent materials, obtained from different possible precursors (e.g., coconut shells, wood, coal), characterized by a wide specific surface area (1200–1300 m^2/g) and microporous volume. These materials are typically employed as adsorbents of ammonia and alcohols for adsorption TES, owing to the high affinity demonstrated towards these adsorbates. Owing to the competitive cost and wide commercial availability, they are considered as a promising option for TES applications. However, they can be employed only for closed adsorption TES, since their affinity towards water vapor is quite limited. Furthermore, there are no examples in the literature of the use of these materials in TES to date. As for the case of silica gels, they have been mainly investigated as possible substrates and matrices for composite adsorbents, primarily exploited to increase the poor thermal conductivity of adsorption materials.

3.1.6. Composite Sorbents

Composite sorbents represent a hybrid method to enhance the sorption ability of materials under the typical working boundary conditions of adsorption TES [13]. Indeed, they are based on the embedding of inorganic salt (e.g., $CaCl_2$, LiCl, LiBr) inside a host porous structure (e.g., silica gel, vermiculite). This concept was invented by the Boreskov Institute of Catalysis, in an attempt to exploit the absorption ability of certain kinds of salts, while avoiding one of their main limitations, that is, the excessive mass transfer limitation induced by the agglomeration of the salt when it is employed in bulk. Indeed, as depicted by the working principle reported in Figure 7, the reaction between salt and adsorbate is always confined inside the pores of the host matrix, this makes the adsorption/desorption more stable and less affected by the adsorbate diffusion phenomena. Furthermore, the proper selection of the salt and the pore size of the host matrix offers the ability to nano-tailor the achievable adsorption properties of the synthesized material. Owing to the wide availability of inorganic salts, capable of reacting with different adsorbates, it is possible to synthesize composite sorbents that can be used with a large number of adsorbates. Typical examples are the well-known Selective Water Sorbents, SWSs, which represent the wider class of composite materials, specifically developed for water adsorption.

The literature is full of several different developed composite sorbents for TES applications. They often demonstrate an attractive performance, in terms of heat storage density, especially on a thermodynamic basis. The main issues displayed by these materials are related to possible salt leakage from pores and slow kinetic behavior, due to the chemical reaction occurring between salt and adsorbate.

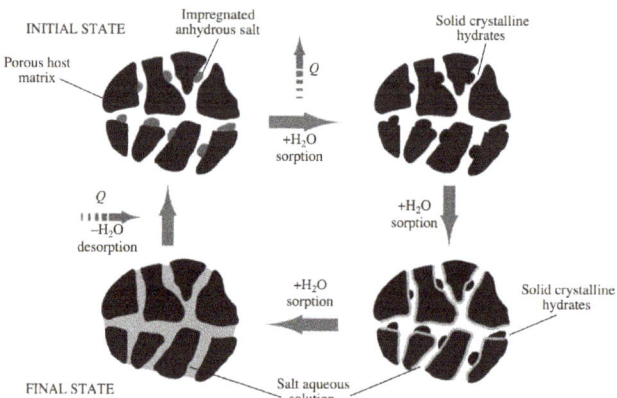

Figure 7. The working principle of a composite adsorbent [13], with permission from Springer Nature, 2007.

3.2. Adsorption Material Heat Storage Calculations

An interesting comparison among the several adsorbent materials introduced above is reported in [14]. The calculations are only performed for closed systems with water as the adsorbate, as these are still the most widely employed adsorption TES under investigation. In the present paragraph, some of the main obtained outcomes are summarized. Generally, in order to calculate the most effective adsorbent material for TES, the main parameter to be investigated is represented by the integral enthalpy of adsorption, which can be easily calculated, as reported by the following equation:

$$H_{ads} = \Delta H_{ads} \left(w_{max} - w_{min} \right) \; [J/g_{ads}] \tag{1}$$

where:

- H_{ads} [J/g_{ads}] is the enthalpy of adsorption, which can be considered as the achievable heat storage density at a material level;
- ΔH_{ads} [J/g_{water}] is the differential enthalpy of adsorption referred to the adsorbed amount of water;
- w_{max} and w_{min} [g_{water}/g_{ads}] are the maximum and minimum adsorption amount of water over the adsorbent material, at the given working boundary conditions.

The value of the differential enthalpy of adsorption is generally calculated through the measurement of the equilibrium adsorption curves according to the well-known Clausius-Clapeyron equation [15]. Authors in [14] demonstrated that SAPO-34 and SWS generally show the highest heat storage capacities regardless of the boundary conditions. Nevertheless, classical zeolites whose integral heat of adsorption is very limited at 90 °C of regeneration temperature, become very attractive when higher temperatures (i.e., 160 °C) are available instead. Silica gels, as discussed previously, maintain quite a limited TES density, which makes this class of adsorbents less attractive for this application.

Finally, it must be emphasized that these calculations have been performed for unit mass. Usually, TES density is calculated on a volumetric basis, since the occupied volume can be an issue, for instance, in the domestic sector. Nevertheless, to calculate the volumetric adsorption storage is quite complicated, since the bulk adsorbent material density strongly depends on grain size and composition. For this reason, in order to compare different adsorbent materials, the gravimetric TES density is taken as a reference parameter. Main features of the introduced sorption material classes are shown in Table 1.

Table 1. Main properties of sorbent material classes.

	Silica Gels	Zeolites	ALPO$_S$/SAPO$_S$	Composites	MOFs	Activated Carbons
Adsorption heat (kJ/kg)	160 ÷ 180 *	50 ÷ 300*	250 ÷ 300 *	50 ÷ 250*	20 ÷ 200 **	45 ÷ 900 ***
Typical desorption temperatures [°C]	50 ÷ 80	70 ÷ 350	60 ÷ 90	60 ÷ 90	60 ÷ 150	80 ÷ 200
Density (kg/m^3)	650 ÷ 700	650 ÷ 900	800 ÷ 900	300 ÷ 600	1000 ÷ 2000	700 ÷ 750
Specific heat (kJ/kgK)	0.8 ÷ 0.9	0.85 ÷ 0.95	0.85 ÷ 0.95	0.95 ÷ 1.05	0.8 ÷ 1.2	0.8 ÷ 1.5
Thermal conductivity (W/mK)	0.15 ÷ 0.20	0.15 ÷ 0.25	0.15 ÷ 0.25	0.15 ÷ 0.30	0.10 ÷ 015	0.15 ÷ 0.75
Possible refrigerants	water	water	water	water, methanol, ethanol	water, methanol, ethanol	methanol, ethanol, ammonia
Amount of uptake exchanged in a typical cyle [g/g]	0.03 ÷ 0.10	up to 0.2	up to 0.25	up to 0.8	0.16 ÷ 0.40	015 ÷ 0.60

* the heat of adsorption is calculated for a cycle with T_{des} = 100 °C, T_{cond} = 30 °C, T_{ads} = 50 °C, T_{ev} = 10 °C, with water as sorbate; ** the heat of adsorption is calcualted from isotherms at 298 K, 303 K and 333 K, with water as sorbate; *** the range of heat of adsorption is calculated with methanol and ammonia as sorbates.

3.3. Adsorption TES Components

The development of closed adsorption TES components is mainly focused on the core component represented by the adsorber unit, realized by putting the adsorbent material in contact with an efficient heat exchanger. Indeed, since the adsorbent materials present a very low thermal conductivity, their inertia towards thermal cycles is quite high. This behavior is reflected in a limited dynamic performance which causes a reduced achievable specific power (both in terms of mass and volume), leading to bulky components. With regards to this, several activities have been carried out in the past to optimize the thermal conductivity of adsorbent materials by adding pieces of highly conductive materials [16]. More recently, research activity has been oriented mainly towards the optimization of the adsorber unit itself, to reduce the heat and mass transfer resistance between the heat exchanger and the adsorbent material. Thus far, three main technologies have been identified to realize effective adsorbers: Loose grains, binder-based coating, and in-situ crystallization coating techniques.

The less expensive and widely employed technique for adsorption TES is that which is based on loose grains embedded in the heat exchanger (HEX). In this case, the main goal is to find the best compromise, in terms of grain size, which allows for a good heat-transfer efficiency without significantly affecting the vapor diffusion through the adsorber.

The binder-based coating technique, is based on the reduction of contact resistance between the heat exchanger and the adsorbent material by distributing a thin, homogeneous layer over the wide heat transfer area of the HEX itself. In this way, heat transfer is enhanced, since the contact between the adsorbent and HEX is uniform and not punctuated, as in the loose grains configuration. Recently, potentialities of coatings on advanced HEX supports (graphite plates) have also been investigated [17]. Experimental results demonstrated that this approach could enhance the kinetic performance of the components, thus increasing the power of the developed units. Nevertheless, some parameters need to be carefully investigated in order to optimize the binder-based coating technique. Indeed, the adsorbent layer thickness needs to be carefully controlled, to avoid excessive mass-transfer resistance and to reach a high mechanical stability level. Furthermore, despite their good mechanical properties, organic binders can release small quantities of non-condensable gases, which can affect the performance of the adsorber itself.

The last option, currently under development, is the in-situ crystallization technique. This is mainly oriented towards zeolite and zeo-like materials, which are crystalline and can be directly synthesized over the metallic substrate of the HEX, leading to a perfect thermal contact, dramatically reducing the heat transfer resistance. This technique has already been applied to full-scale adsorbers and confirmed to be very promising from a dynamics point of view [18]. The main limitations, still under investigation, are related to the long duration and high energy consumption of the crystallization process, and to the low amount of adsorbent that can be deposited over the HEX, which can affect the achievable volumetric power.

3.4. Adsorption TES Systems

Adsorption TES systems are still in the early stages of development and are not yet completely commercialized. Nevertheless, some particular applications have already been put on the market because they fit certain needs perfectly. In this paragraph a brief collection of recently developed adsorption TES applications is reported.

A classic example is the self-cooling portable beer barrel developed by ZeoTech company [19]. This system perfectly exploits the peculiarity of the adsorption TES. Indeed, it consists of a zeolite embedded inside the external shell of the barrel, kept separate from the evaporator (Figure 8). Once the beer needs to be cooled down, the connection between the anhydrous zeolite and the evaporator is opened through a manual valve, and the heat of evaporation is subtracted from the beer, which is cooled down to the desired temperature. When the barrel is empty, the saturated zeolite is regenerated in the oven, thus performing the charging phase. This process perfectly applies adsorption heat storage to small-scale apparatuses.

Figure 8. A small beer barrel with embedded adsorption TES for cooling. Courtesy of Cool-System (figure adapted from [19] with permission of Springer, 2007).

A different application of adsorption TES was recently developed and commercialized by Bosch in collaboration with ZAE Bayern, in which an open adsorption TES has been optimized to enhance the energetic performance of a dishwasher (Figure 9). The working principle can be found elsewhere [20]. Essentially, in this application, the thermal energy is reversibly stored in a zeolite cartridge, which is regenerated (charged) in the first phase of the washing process and discharged, with the release of a large amount of heat, during the drying phase. In this way, it becomes possible to reduce the amount of electrical energy required to perform an entire washing cycle, enhancing the energetic class of the appliance.

Another reported example of adsorption TES, is a large-scale system for industrial heat recovery, storage, and transportation, based on an open adsorption cycle. Figure 10 summarizes this concept, developed at ZAE Bayern laboratories [21]. It consists of recovering heat from an industrial site, by flowing hot air through a zeolite 13X bed. Once the adsorbent material is regenerated, the reactor full of dried zeolite (charged TES) is transported to the site, where it is discharged by flowing humid air through the zeolite bed, thus releasing heat to drive another industrial process. This system proved to be quite promising as the demand and the user side are closely related, as it is well known that industrial sites are one of the major sources of waste heat worldwide [22]. Therefore, to increase the share of this type of application, it is necessary to carefully analyze the boundary conditions, in order to make the process economically feasible.

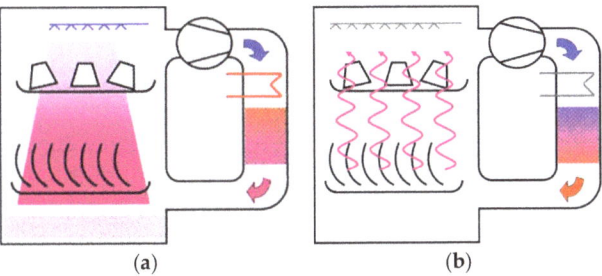

Figure 9. An adsorption based dishwasher. Washing phase (**a**); drying phase (**b**).

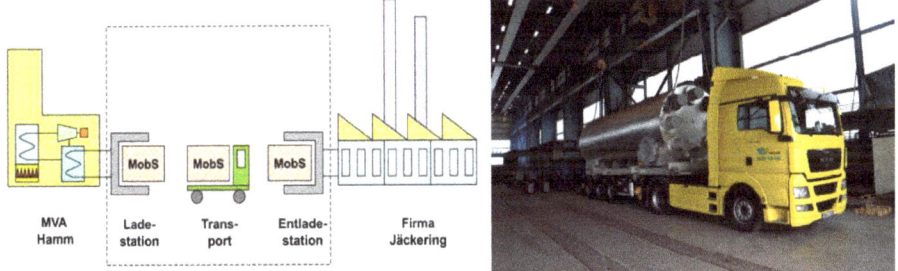

Figure 10. Mobile adsorption TES for industrial applications.

The final example is a prototype of a compact TES for mobile applications [23]. The sorption TES has been realized and tested at CNR-ITAE. It consists of two vacuum chambers, for the adsorber and the phase changer (Figure 11). Certain flanges allow connection to the sensors for monitoring of the most relevant parameters (pressure, temperature) and the other components of the device. The exchanger used for the adsorber is a flat-tube and fins-type with an exchange area of 1.75 m². The exchanger has been filled with 4.3 kg of AQSOA™-Z02 grains, in the range 1–2 mm. The material has then been contained by means of a metallic net. The connection between the adsorber and the phase changer is realized through an electrically actuated pneumatic valve.

The phase changer consists of a welded chamber, containing four high efficiency fin-and-tube heat exchangers with copper fins and stainless steel tubes connected in parallel through an external tubular steel manifold. Each one has an exchange surface of 1.33 m². Vacuum flanges allow connections to the adsorber and to some sensors.

The test set-up is completed by a hydraulic circuit realized with copper Φ12 mm tubes, thermally insulated by a polyurethane foam with a thickness of 1.5 cm. The hydraulic circuits include four 3-way valves.

The results obtained with this system showed good performance and efficiency, consistent with those recently measured by others and theoretically predicted [24,25]. In particular, storage capacities up to 263 Wh/kg of adsorbent were obtained, corresponding to a storage capacity 40% higher than that of water under the same boundary conditions. A peculiarity of the investigated system is the possibility to use low-temperature waste heat (T < 100 °C) both for heat and cold storage purposes: the measurements carried out highlighted that even with a heat source temperature at 85 °C, temperatures of 5–10 °C can be efficiently produced.

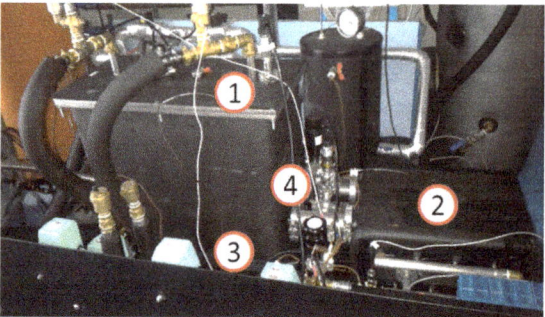

Figure 11. The TES prototype system realized at CNR-ITAE. 1: Adsorber, 2: Phase changer, 3: Hydraulics. 4: Vacuum valve.

4. Conclusions and Future Perspectives

The present paper has summarized some of the main features of the adsorption TES, analyzing its state of development from different points of view, from the material up to the system applications. Research activity is still ongoing, in an attempt to solve the main issues related to this technology. Particularly, the following ways seem to be quite promising in order to reach commercial diffusion during next years:

- At a material level, the main challenge is to reduce the costs of available materials, in order to make the adsorption TES systems more competitive. In this context, a lot of efforts are put into the employment of less expensive raw materials for zeo-like adsorbents and to reduce the hydrophilicity of classical zeolites, to keep down the required regeneration temperature. Furthermore, MOFs are continuously under development, thanks to their promising features.
- At a component level, the main task is the realization of efficient adsorbers, based either on adsorbent coating or on a loose grains technique, which can allow the size of the storage systems to be limited, enhancing the kinetic performance. Furthermore, particular attention is also placed on the reliability of these components, in terms of corrosion and hydro-thermal stability issues.
- At a system level, small-scale adsorption TES units are under development for domestic applications. Indeed, if properly coupled to the distribution system, they can not only store thermal energy, but also provide a heat-pumping effect during the winter season and cooling energy during the summer season, thus making this a component for fully-integrated heat and cold storage, throughout the year.

Funding: This research received no external funding.

Conflicts of Interest: The authors declare no conflict of interest.

References

1. Alva, G.; Lin, Y.; Fang, G. An overview of thermal energy storage systems. *Energy* **2018**, *144*, 341–378. [CrossRef]
2. Velraj, R. Sensible heat storage for solar heating and cooling systems. In *Advances in Solar Heating and Cooling*; Elsevier: New York, NY, USA, 2016; pp. 399–428.
3. Mehling, H.; Cabeza, L.F. *Heat and Cold Storage with PCM—An Up to Date Introduction into Basics and Applications*; Springer: Berlin, Germany, 2008.
4. Aneke, M.; Wang, M. Energy storage technologies and real life applications—A state of the art review. *Appl. Energy* **2016**, *179*, 350–377. [CrossRef]

5. Pfleger, N.; Bauer, T.; Martin, C.; Eck, M.; Wörner, A. Thermal energy storage—Overview and specific insight into nitrate salts for sensible and latent heat storage. *Beilstein J. Nanotechnol.* **2015**, *6*, 1487–1497. [CrossRef] [PubMed]
6. Hauer, A. Sorption Theory for Thermal Energy Storage. In *Thermal Energy Storage for Sustainable Energy Consumption*; Paksoy, H.O., Ed.; NATO Science Series; Springer: Dordrecht, The Netherlands, 2007; Volume 234, pp. 393–408.
7. Bales, C. Laboratory Tests of Chemical Reactions and Prototype Sorption Storage Units. IEA-SHC Task 32 Report B4 of Subtask B Chemical and Sorption Storage. Available online: http://members.iea-shc.org/publications/downloads/task32-b4.pdf (accessed on 15 March 2016).
8. Jaehnig, D.; Hausner, R.; Wagner, W.; Isaksson, C. Thermo-chemical storage for solar space heating in single-family house. In Proceedings of the 10th International Conference on Thermal Energy Storage, Atlantic City, NJ, USA, 31 May–2 June 2006.
9. Yu, N.; Wang, R.Z.; Wang, L.W. Sorption thermal storage for solar energy. *Prog. Energy Combust. Sci.* **2013**, *39*, 489–514. [CrossRef]
10. Jänchen, J.; Stach, H. Adsorption properties of porous materials for solar thermal energy storage and heat pump applications. *Energy Procedia* **2012**, *30*, 289–293. [CrossRef]
11. ZEOLITE AQSOA | Products | Mitsubishi Chemical Corporation. Available online: https://www.m-chemical.co.jp/en/products/departments/mcc/aquachem/product/1201185_8078.html (accessed on 7 July 2018).
12. Rezk, A.; Al-Dadah, R.; Mahmoud, S.; Elsayed, A. Characterization of metal organic frameworks for adsorption cooling. *Int. J. Heat Mass Transf.* **2012**, *55*, 7366–7374. [CrossRef]
13. Aristov, Y.I. New family of solid sorbents for adsorptive cooling: Material scientist approach. *J. Eng. Thermophys.* **2007**, *16*, 63–72. [CrossRef]
14. Bales, C. Thermal Properties of Materials for Thermo-Chemical Storage of Solar Heat. IEA-SHC Task 32 Report B2 of Subtask B "Chemical and Sorption Storage". Available online: http://members.iea-shc.org/publications/downloads/task32-Thermal_Properties_of_Materials.pdf (accessed on 15 March 2016).
15. Talu, O.; Kabel, R.L. Isosteric heat of adsorption and the vacancy solution model. *AIChE J.* **1987**, *33*, 510–514. [CrossRef]
16. Demir, H.; Mobedi, M.; Ülkü, S. A review on adsorption heat pump: Problems and solutions. *Renew. Sustain. Energy Rev.* **2008**, *12*, 2381–2403. [CrossRef]
17. Vasta, S.; Giacoppo, G.; Barbera, O.; Calabrese, L.; Bonaccorsi, L.; Freni, A. Innovative zeolite coatings on graphite plates for advanced adsorbers. *Appl. Therm. Eng.* **2014**, *72*, 153–159. [CrossRef]
18. Bauer, J.; Herrmann, R.; Mittelbach, W.; Schwieger, W. Zeolite/aluminum composite adsorbents for application in adsorption refrigeration. *Int. J. Energy Res.* **2009**, *33*, 1233–1249. [CrossRef]
19. Hauer, A. Adsorption Systems for TES—Design and Demonstration Projects. In *Thermal Energy Storage for Sustainable Energy Consumption*; Paksoy, H.O., Ed.; NATO Science Series; Springer: Dordrecht, The Netherlands, 2007; Volume 234, pp. 409–428.
20. Hauer, A.; Fischer, F. Open Adsorption System for an Energy Efficient Dishwasher. *Chem. Ing. Tech.* **2011**, *83*, 61–66. [CrossRef]
21. Krönauer, A.; Lävemann, E.; Brückner, S.; Hauer, A. Mobile Sorption Heat Storage in Industrial Waste Heat Recovery. *Energy Procedia* **2015**, *73*, 272–280. [CrossRef]
22. Miró, L.; Brückner, S.; Cabeza, L.F. Mapping and discussing Industrial Waste Heat (IWH) potentials for different countries. *Renew. Sustain. Energy Rev.* **2015**, *51*, 847–855. [CrossRef]
23. Palomba, V.; Vasta, S.; Freni, A. Experimental testing of AQSOA FAM Z02/water adsorption system for heat and cold storage. *Appl. Therm. Eng.* **2017**, *124*, 967–974. [CrossRef]
24. Van Alebeek, R.; Scapino, L.; Beving, M.A.J.M.; Gaeini, M.; Rindt, C.C.M.; Zondaga, H.A. Investigation of a household-scale open sorption energy storage system based on the zeolite 13X/water reacting pair. *Appl. Therm. Eng.* **2018**, *139*, 325–333. [CrossRef]
25. Girnikab, I.S.; Aristov, Y.I. Dynamic optimization of adsorptive chillers: The "AQSOA™-FAM-Z02–Water" working pair. *Energy* **2016**, *106*, 13–22. [CrossRef]

© 2018 by the authors. Licensee MDPI, Basel, Switzerland. This article is an open access article distributed under the terms and conditions of the Creative Commons Attribution (CC BY) license (http://creativecommons.org/licenses/by/4.0/).

Article

New Composite Water Sorbents CaCl₂-PHTS for Low-Temperature Sorption Heat Storage: Determination of Structural Properties

Alenka Ristić [1,*] and Nataša Zabukovec Logar [1,2]

1. Department of Inorganic Chemistry and Technology, National Institute of Chemistry Slovenia, Hajdrihova 19, SI-1001 Ljubljana, Slovenia; natasa.zabukovec@ki.si
2. School of Science, University of Nova Gorica, Vipavska 13, 5000 Nova Gorica, Slovenia
* Correspondence: alenka.ristic@ki.si; Tel.: +386-1-47-60-215

Received: 16 November 2018; Accepted: 20 December 2018; Published: 26 December 2018

Abstract: Sorption heat storage, as one of low-energy consuming technologies, is an approach to reduce CO_2 emissions. The efficiency of such technology is governed by the performance of the applied sorbents. Thus, sorbents with high water sorption capacity and regeneration temperature from 80 to 150 °C are required. Incorporation of hygroscopic salt such as calcium chloride into porous materials is a logical strategy for increasing the water sorption capacity. This work reports the study on the development of composites with PHTS (plugged hexagonal templated silicate) matrix with an average pore size of 5.7 nm and different amounts of calcium chloride (4, 10, 20 wt.%) for solar thermal energy storage. These composites were prepared by wetness incipient impregnation method. Structural properties were determined by X-ray diffraction (XRD), nitrogen physisorption, scanning electron microscopy (SEM) and transmission electron microscopy (TEM). $CaCl_2$ was confined in micro- and mesopores of the matrix. The resulting $CaCl_2$-PHTS materials were used for water sorption at 40 °C, showing an increase of maximal water uptake with higher amount of calcium chloride from 0.78 g/g to 2.44 g/g of the dry composite. A small reduction in water uptake was observed after 20 cycles of sorption/desorption between temperatures of 140 °C and 40 °C, indicating good cycling stability of these composites under the working conditions.

Keywords: $CaCl_2$-PHTS; composites; water sorption; heat storage; structural properties

1. Introduction

Thermal energy storage (TES) is becoming a crucial technology in enabling more efficient use of renewable energy and contributing to the reduction of our dependency on fossil fuels. It can be divided into three main categories according to how energy is stored: sensible heat (e.g., water tanks, underground storage), latent heat (e.g., ice, phase change materials), and thermochemical heat storage [1]. Thermochemical heat storage utilizes the reversible chemical reaction [2] and/or sorption processes [3] involving working fluids and solids or liquids. Sorption thermal energy storage depends on thermo-physical properties of the sorbents such as thermally stable microporous and mesoporous materials. The main criteria for the selection of a proper sorbent for sorption thermal energy storage are high sorption capacity, low desorption temperature, and high temperature level of released heat of adsorption [4]. A large number of sorbents are currently considered for sorption thermal energy storage, traditional such as silica gels and zeolites, and innovative like aluminophosphates, MOFs, and composites [5,6]. The most versatile class of sorbents are the two-component sorbents or composite salt in porous inorganic matrix (CSPM) [7] which combine the advantages of the pure porous matrix and hygroscopic salt hydrates for the enhancement of water sorption capacity, heat, and mass transfer on one side, and on the other hand to avoid or reduce the deliquescence and agglomeration of salt hydrates during sorption/desorption cycles. The composites developed thus far can be used also with other working fluids, such as methanol [8]. The sorption properties of the composites can be tailored by varying chemical nature, amount, and particle size of the incorporated salt and depend strongly on the structural and physico-chemical properties of the porous host matrix (pore size/shape, pore volume, and hydrophilic properties) and host-salt interactions. Further advantages are a low desorption temperature, a low price and a simple preparation method [9]. Besides wet impregnation and incipient wetness impregnation procedures, microencapsulation is another approach for the stabilization of salt hydrates by enveloping with a second inherently stable material to prevent coalescence or agglomeration [10].

As a safe, environment-friendly, and available sorbate, water is usually the preferred choice. A combination of three mechanisms, adsorption by the host matrix, chemical reaction between water and salt, and absorption by the salt aqueous solution in the pores, determines water sorption behavior of the composites [11]. Typical hygroscopic salt hydrates incorporated in the composites for sorption TES are LiCl, $MgCl_2$, $CaCl_2$, $MgSO_4$, and $SrBr_2$ [1]. $CaCl_2$ has been combined several times with different inorganic host materials such as silica gels [12], disordered mesoporous iron silicate [13], ordered mesoporous silicas [14], mesoporous alumina-silica [15], Ca-exchanged binder-free zeolite X [16], alumina [17], clays [18], MOFs [19], and carbonaceous structures [20] because of its low cost, non-toxicity, large availability, and a high sorption capacity [21]. In general, the role of the matrix is to adsorb water and serve as a dispersion medium, which forms a required salt particle size and high salt surface area. In addition, this prevents forming of typical agglomeration of salt particles and conducts heat through the solid. Usually, the porous matrix has lower water sorption capacity than the hygroscopic salts, which interact with water to increase the sorption capacity [11].

Mesoporous ordered silicates, MCM-41 and SBA-15, are a special class of materials that possess pores with diameters in the range of 2 to 50 nm and can adsorb large amount of water due to the amorphous surface structure and high pore volumes, exhibiting water adsorption isotherms of Type V according to the UIPAC classification [22]. These materials possess pore sizes from 2 to 15 nm, large surface areas (>500 m^2/g), excellent thermal stability, and hexagonal pore arrangement. A large amount of water can be sorbed on them followed by capillary condensation. Basically, all the materials show high water sorption due to their mesoporosity, which is larger than that of the zeolites. The structural characteristics of ordered mesoporous matrices, such as specific surface area, the pore volume, and pore size, determine the sorption properties of the composite sorbent. The pore structure influences on the strength of the interactions between water molecules and the adsorbent sorption sites. Higher surface area means more available sorption sites in the material and indicates better diffusivity of the vapor, which is crucial for optimal mass and heat transfer. Maaz et al. [23] stated that synthesis and post-synthesis procedures could have a great influence on water sorption behavior of SBA-15, correlating mainly with the amount and type of silanol groups on the silica surface and increasing the water uptake over a broad range of relative pressure due to high micropore volume and high surface polarity. Only a few investigations have been performed on the incorporation of $CaCl_2$ into ordered mesoporous silicates with mono-sized pores such as MCM-41 [24] and SBA-15 [11,25,26]. These matrices with uniform pore dimension provide an effective tool for controllable tuning of the solvation temperature of the confined salt. This temperature is higher when the salt is located in smaller pores [27]. Ponomarenko et al. prepared a composite of SBA-15 (pore diameter 7.5 nm) and 43 wt. % of $CaCl_2$ by wet impregnation, showing on filling the pores with $CaCl_2$ and deposition of the salt on the surface [14]. The authors showed that the Type IV nitrogen sorption isotherm of pure SBA-15 was changed to Type III, which is typical for nonporous adsorbent, indicating the collapse of the ordered matrix mesopore structure after the impregnation [28]. Ristić et al. prepared the composite combining SBA-15 matrix (average mesopore size of 10.2 nm) and 4 wt.% $CaCl_2$ by wet impregnation. A low content of salt was used in order to maintain the ordered mesostructure [26]. The composite was exposed to a short-cycle hydrothermal treatment consisting of 20 cycles between temperatures of 150 °C and 40 °C at a water vapor pressure of 56 mbar, showing good initial hydrothermal stability of the composites under the operating conditions. In addition, Glaznev et al. prepared two SBA-15/CaCl$_2$ composites with average mesopores: 8.5 and 11.8 nm and containing 28.2 wt.% and 29.5 wt.% of the salt, respectively. It was shown that changing the mesopore size of the matrix can influence the vapor transport [25]. Smaller salt particles in smaller mesopores sorbed water easier. The aim of both studies were to prepare the composites with the highest possible amounts of CaCl$_2$ to achieve the highest energy storage capacities without detailed investigations on the structural properties and stability of these composites during water sorption, which indeed cause structural modifications.

Plugged hexagonal templated silicate (PHTS) has the same hexagonal pore arrangement as SBA-15, however, some of its cylindrical mesopores have internal porous plugs, while others are open without any major constrictions [29]. They are prepared by a modification of synthesis procedure for SBA-15, such as changing temperature and time of aging or molar ratio of reactants. The hydrothermal stability of PHTS under steaming was found to be better than that of SBA-15. In comparison to SBA-15, the micropore volume of PHTSs tend to be enhanced, hence, it can be envisioned that some of these micropores are located in the plugs. Thus, the mesoporous structure of PHTS features compartments, which appear to be accessible through the voids between constrictions or pores in the plugs [30].

Here, we present a study on the influence of water sorption on the structural properties of the composites of the plugged hexagonal templated silica (PHTS) with different contents of calcium chloride. The effect of CaCl$_2$ amount on the PHTS matrix and on the water sorption capacity of composites is investigated. Water sorption properties of the composite were studied for low-temperature heat storage application, while stability of these composites was tested during 20 cycles between 40 °C and 140 °C.

2. Materials and Methods

2.1. Materials

PHTS was synthesized by modification of the SBA-15 synthesis using long-chain surfactant triblock copolymer Pluronic P123 [31] to prepare the plugged hexagonal templated silica with ordered hexagonal mesopore arrangement with average pore size lower than 6 nm. In a typical synthesis procedure, Pluronic P123 was dissolved in diluted HCl solution by stirring at room temperature until P123 was dissolved. Then tetraethylortho silicate was added, and the resulting mixture was stirred for 8 h at 65 °C, and then kept at 65 °C at ambient pressure for 16 hours without stirring. The solid product was filtered and washed repeatedly with deionized water. After drying at room temperature overnight, the product was calcined in air at 550 °C for 6 h in order to remove the surfactant. The composites were prepared by incipient wetness impregnation [32] of the PHTS matrix with concentrations: 4 wt.%, 10 wt.%, and 20 wt.% of calcium chloride. The samples were dried at room temperature overnight. The composites were denoted 4-CaCl$_2$-PHTS, 10-CaCl$_2$-PHTS and 20-CaCl$_2$-PHTS.

2.2. Methods

The X-ray powder diffraction (XRPD) patterns were recorded on a PANalytical X'Pert PRO high-resolution diffractometer (Almelo, The Netherlands) with Alpha1 configuration using CuK$_\alpha$1 radiation (1.5406 Å) in the range from 0.5 to 35° 2θ with step 0.017° per 100 s using a fully opened X'Celerator detector. Morphology of the matrix and the composites was studied by scanning electron microscopy (SEM) on a Zeiss SupraTM 3VP SEM microscope (Jena, Germany). Transmission electron microscopy (TEM) micrographs were obtained on a 200-kV field-emission gun (FEG) microscope JEOL JEM 2010F (Peabody, MA, USA). Elemental analysis was performed by energy dispersiveX-ray analysis (EDAX) with an INCA Energy system attached to a Zeiss SupraTM 3VP microscope (Jena, Germany). Nitrogen physisorption measurements were performed at −196 °C on a Tristar volumetric adsorption analyzer (Micromeritics, Norcross, GA, USA). Before the adsorption analysis, the samples were outgassed under vacuum for 2 h at 200 °C in the port of the adsorption analyzer. Prior to the evaluation of textural properties of the composites the amount of the nonporous salt was taking into account which does not contribute to nitrogen adsorption to a large extent. Thus nitrogen isotherms and all specific values (surface area, pore volume) were corrected. The BET specific surface area [33], S_{BET}, was calculated using the adsorption branch in the relative pressure range between 0.05 and 0.16. The total pore volume, Vt, was estimated from the amount adsorbed at a relative pressure of 0.96. The pore size distributions (PSDs) were calculated from nitrogen adsorption data using an algorithm based on ideas of Barrett, Joyner, and Halenda (BJH) [34]. The maxima on the PSD are considered as the primary mesopore diameters for given samples. Water sorption characteristics of the samples were determined by an IGA-100 gravimetric analyzer (Hiden Isochema Ltd., Warrington, UK). Water sorption isotherms were obtained at 25 and 40 °C by setting equal pressure intervals of 1.6 mbar between vacuum and 40 mbar (saturation vapor pressure of 73.8 mbar at 40 °C) with an equilibrium time of 80 s. Before adsorption measurements, the samples were outgassed to a constant weight under ultrahigh vacuum (<10^{-5} mbar) at 150 °C for 5 hours. The hydrothermal stability of the materials was evaluated with 20 cycle measurements in a helium gas flow with 75% relative humidity by varying the temperature between 40 and 140 °C at 56 mbar. The relative humidity was controlled by varying the ratio of dry and saturated helium via two mass flow controllers. The water capacity of materials was measured at the beginning and after the 20 cycles. The definition of the thermodynamic heat cycle and the calculation of the amount of heat involved are given by De Lange et al. [35]. First, the adsorption equilibrium data of each sample obtained using the IGA-100 was plotted as the so-called characteristic curve (adsorbed water uptake as a function of the adsorption potential A). The adsorption potential A is defined as: $A = RT\ln(p_s(T))/p$ (1), where R is the gas constant, T the temperature, p_s the saturation pressure, and p the vapor partial pressure [36]. The integral enthalpy of adsorption Q_{ads} can be calculated by the following equation: $Q_{ads} = \Delta H_{ads} (w_{ads} - w_{des})$ [kJ/kg$_{ads}$] (2), where Q_{ads} [kJ/kg$_{ads}$] is the enthalpy of adsorption, which can be considered as the achievable heat storage density at a material level; ΔH_{ads} [kJ/kg$_{water}$] is the differential enthalpy of adsorption referred to the adsorbed amount of water; w_{ads} and w_{des} [kg$_{water}$/kg$_{ads}$] are the maximum and minimum adsorption amount of water over the adsorbent material at the given boundary conditions [6]. As the specific heat of the adsorbents is not known exactly, it was set to 1 kJ/kg K. The results were not affected by this value in a significant way [37]. The value of the differential enthalpy of adsorption was calculated through the measurement of the equilibrium adsorption curves according to the well-known Clausius-Clapeyron equation. The integral heat of adsorption was calculated for heat storage application according the literature: at a desorption temperature of 120 °C, which can be attained by solar thermal collectors. The sorption temperature was fixed to 40 °C, which is sufficient for space heating applications. The water vapor pressure during desorption and adsorption of the samples was set to 12.3 mbar (a dew point temperature of 10 °C). The difference of adsorbed water amount at 40 °C and 120 °C at 12.5 mbar is the cycle (water) loading lift of the composite.

3. Results and Discussion

3.1. Structural Properties of As-Prepared Smples

The synthesis procedure of the SBA-15 was modified in order to synthesize the PHTS matrix with average pore size of 5.7 nm with the aim to confine $CaCl_2$ in the pores of this matrix. PHTS was synthesized at ambient pressure at 65 °C, while SBA-15 synthesis involved hydrothermal treatment at 100 °C [38]. It is known that aging temperature and time affects the pore size of the SBA-15, namely a higher aging temperature, increases average pore size [23]. Wetness incipient (dry) impregnation was used for the preparation of the composites.

Figure 1a shows low-angle X-ray powder diffraction patterns of the PHTS and the composites containing different amounts (4, 10, and 20 wt.%) of calcium chloride. The PHTS matrix pattern illustrates three diffraction peaks corresponding to the reflections typical for two-dimensional (2D)-hexagonal pore arrangement. It can be seen that after loading of 4 wt.% of calcium chloride, three diffraction peaks were still present and their 2θ values were only slightly shifted, indicating that the channels with good order were maintained during the preparation procedure of the composite. The impregnation of larger amounts of calcium chloride of the matrix leads to the change of the diffraction patterns; only one less intensive diffraction peak was observed for the composite with 10 wt.% of the salt, while a broad diffraction peak can be seen for the composite with 20 wt.% of the salt. These results indicate the collapse of ordered arrangement of mesopores of these composites into disordered mesopore arrangement. The X-ray diffractograms recorded in the wide-angle range ($5° < 2θ < 35°$) are displayed in Figure 1b. XRD pattern of the composite with 4 wt.% and 10 wt.% of calcium chloride did not show any reflections of calcium chloride, which could be explained with the presence of highly dispersed calcium chloride with nanosized dimensions that are located on the surface and within the pores. Only one broad peak was observed in the range ($15° < 2θ < 30°$) corresponding to glass-like amorphous silicate particles. While the XRD pattern of the composite with 20 wt.% of calcium chloride showed diffraction peaks of the salt.

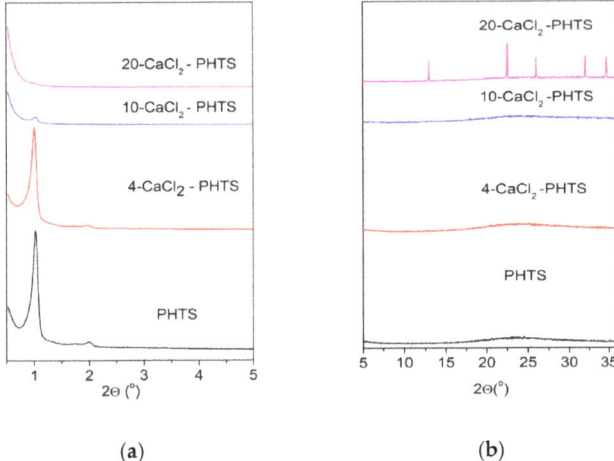

Figure 1. (a) Low-angle XRD patterns of the matrix and the composites containing different contents of $CaCl_2$; (b) High-angle XRD patterns of the pure matrix and the composites with different amounts of $CaCl_2$.

The SEM image of the matrix presented in Figure 2a shows curved rod-like aggregates of matrix particles with a relatively smooth surface. The SEM images of the composite samples with 4 wt.% and 10 wt.% $CaCl_2$ represented in Figure 2b,c, respectively, show similar morphology of the particles without any changes after the loading of calcium chloride solution. There are no observable changes of the outer surface of the composite particles. On the other hand, Figure 2d clearly shows different outer surface of the composite due to higher amount of salt (20 wt.%), which is in accordance with the high-angle XRD pattern (Figure 1b) of this sample.

Pore arrangement of mesoporous materials in local scale was investigated by using transmission electron microscopy (TEM). TEM images (Figure 3) revealed the ordered hexagonal pore arrangement of the PHTS matrix and of the 4-$CaCl_2$-PHTS, while the composites with 10 and 20 wt.% of the salt showed the presence of the disordered mesostructured. A larger amount (20 wt.%) of the salt led to the formation of disordered mesoporous structure of the composites, as can be seen in low-angle XRD pattern.

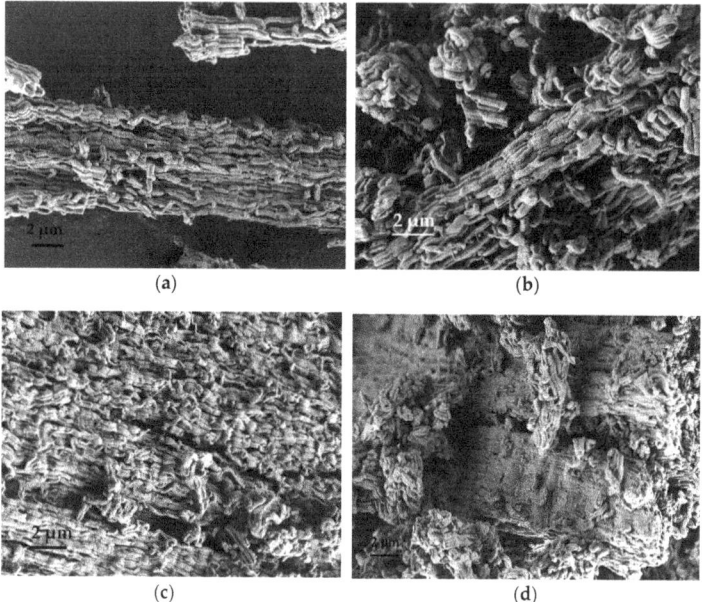

Figure 2. Scanning electron microscopy (SEM) images of (**a**) plugged hexagonal templated silicate (PHTS) matrix; (**b**) 4-CaCl$_2$-PHTS; (**c**) 10-CaCl$_2$-PHTS and (**d**) 20-CaCl$_2$- PHTS.

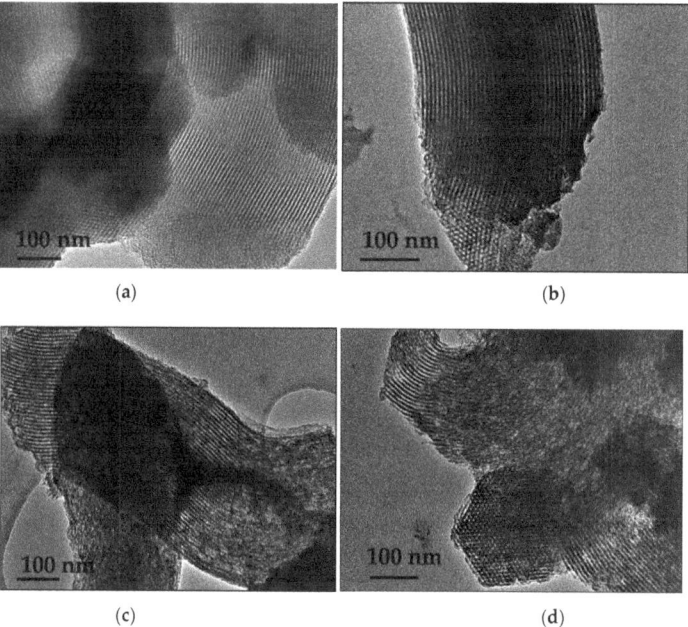

Figure 3. Transmission electron microscopy (TEM) images of the as-prepared (**a**) PHTS matrix, (**b**) 4-CaCl$_2$-PHTS, (**c**) 10-CaCl$_2$-PHTS and (**d**) 20-CaCl$_2$-PHTS.

Porous structure of the PHTS matrix and all composites were examined by nitrogen sorption isotherms. Nitrogen sorption isotherms for PHTS and $CaCl_2$-PHTS are shown in Figure 4a, whereas structural parameters determined on the basis of these isotherms are listed in Table 1. Due to the synthesis procedure PHTS sample exhibits sorption isotherm typical for PHTS-like material [30]. Plugged hexagonal templated silica has the same 2D hexagonal symmetry as SBA-15 with some of its cylindrical mesopores have internal plugs, while others are open as inferred from gas adsorption-desorption data. N_2-sorption isotherms of PHTS are of type IV according to the IUPAC classification with H5 hysteresis loop [28], showing a one-step capillary condensation, two-step desorption, and an appreciable widening of hysteresis loops. The first step is similar to desorption in pure SBA-15 and is assigned to the desorption of N_2 from the open pores; the second desorption step can be attributed to the nanoparticles (plugs) within the mesopores (the narrowed mesopores) [39]. The second step on the desorption branch indicates the existence of plugged mesopores. The presence of $CaCl_2$ in PHTS leads to a marked change in the shape of the hysteresis loops, showing on partial deformation of ordered pore structure (10-$CaCl_2$) into a collapse of the ordered pore arrangement (20-$CaCl_2$), which finds some evidence in TEM and XRD analyses [40]. Additional hysteresis loop at relative pressure above 0.97 on sorption isotherm is present in 10-$CaCl_2$ sample, evidencing the presence of an interparticle or textural porosity [41]. The increase of the amount of the impregnated salt on the matrix leads to a decreased specific surface area, total pore volume, and micropore volume. The decrease of specific surface area is related to the blockage of the smallest pores induced by $CaCl_2$ impregnation. It can be concluded that $CaCl_2$ nanoparticles have been dispersed inside of the micropores and mesopores of the support. Pore size distribution of $CaCl_2$-PHTS materials has been determined using the BJH model, widely used for this type of samples [42]. Although this model often underestimates pore sizes [43], it is appropriate for comparative purposes. Figure 4b displays the pore size distribution determined from adsorption isotherms. As can be observed, the maximum characteristic to open mesopores of PHTS is the most intense and shows an average pore diameter of 5.7 nm.

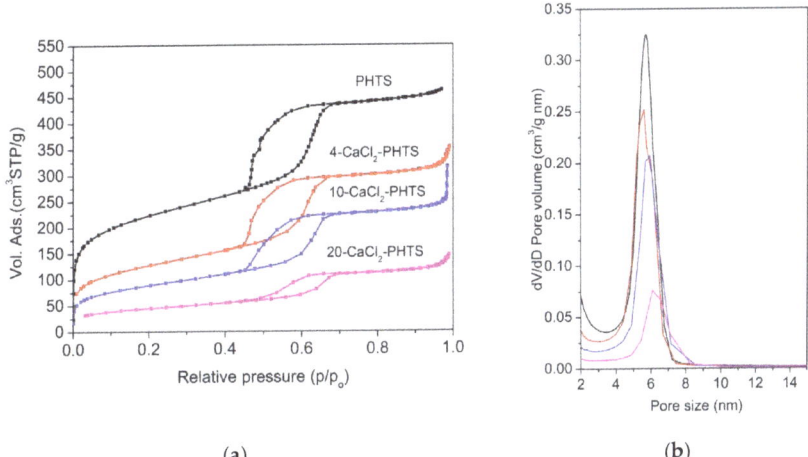

Figure 4. (a) Nitrogen sorption isotherms and (b) pore size distribution of PHTS and composites.

The maximum characteristic for 4-$CaCl_2$-PHTS due to $CaCl_2$ nanoparticles shifted to a lower pore size value (5.6 nm), while in 10-$CaCl_2$ and 20-$CaCl_2$ composites they shift to 5.8 nm and 6.2 nm, respectively. The reason could be the partial destruction of mesoporous structure caused by the corrosive action of calcium chloride solution [44]. Maxima are less intensive with higher amounts of $CaCl_2$.

Table 1. Structural properties of PHTS and the prepared composites.

Sample	S_{BET} (m²/g)	V_{tot} (cm³/g)	V_{mi} (cm³/g)	Average Pore Size (nm)
PHTS	810	0.705	0.122	5.7
4-CaCl$_2$-PHTS	461	0.492	0.037	5.6
10-CaCl$_2$-PHTS	322	0.377	0.022	5.8
20-CaCl$_2$-PHTS	163	0.189	0.016	6.2

Abbreviations: S_{BET}, the BET surface area; V_{tot}, total pore volume evaluated from adsorption isotherm at the relative pressure 0.96.

3.2. Structural Properties of the Samples After Water Sorption and Cycling Test

After the measurement of water isotherms at 40 °C the low-angle XRD patterns (Figure 5) are changed, showing broader less intensive diffraction peaks. The ordered mesoporous structure of the matrix and 4-CaCl$_2$ sample has been retained, as well as the disordered mesoporous structures of 10- CaCl$_2$ and nonporous structure for 20-CaCl$_2$ composite. High-angle XRD patterns of all samples are the same without any diffraction peaks of the salt. It seems that the salt in 20-CaCl$_2$-PHTS was re-dispersed in the PHTS matrix.

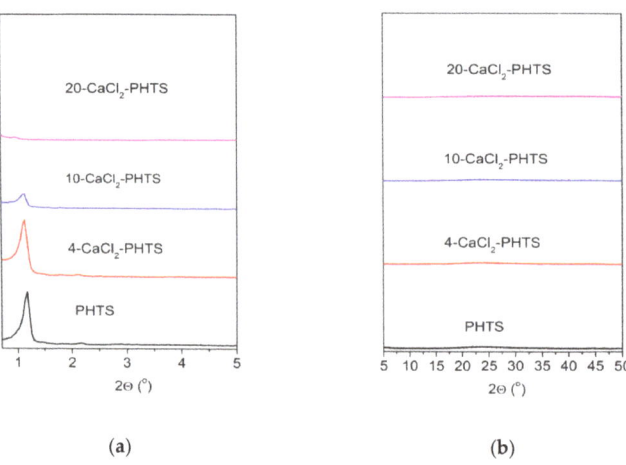

Figure 5. (a) Low-angle XRD patterns of the matrix and the composites containing different contents of CaCl$_2$ after water sorption; (b) High-angle XRD patterns of the pure PHTS and the composites with different amounts of CaCl$_2$ after water sorption.

The shape of nitrogen sorption isotherms (Figure 6) of the composites has also changed, i.e., the hysteresis loops are widening and tailing, exhibiting different types from H2 to H3 [28]. For example, the H2 hysteresis loop is related with pore blocking in silicas after hydrothermal treatment, while the H3 loop is typical of materials with slit-like pores. All isotherms show the presence of interparticle porosity. Nitrogen sorption isotherm of 20-CaCl$_2$ indicates on nonporous material, which is in accordance with XRD. Structural properties are shown in Table 2. Specific surface area and total pore volume decreased after the water sorption measurement. Pore size distributions (Figure 6b) of all samples were broader. The decrease of pore sizes was observed for PHTS, 10-CaCl$_2$ and 20-CaCl$_2$, while significant increase in pore size was seen for 4-CaCl$_2$. It may be due to the corrosion of the walls due to the salt, confined in the intra-walled pores, which interconnect the channels [45] and form larger pores. It can be concluded that the confinement of the salt was not permanent and after hydration and dehydration the salt was re-dispersed, which caused further blocking of pores (20-CaCl$_2$) due to a possible agglomeration of the salt.

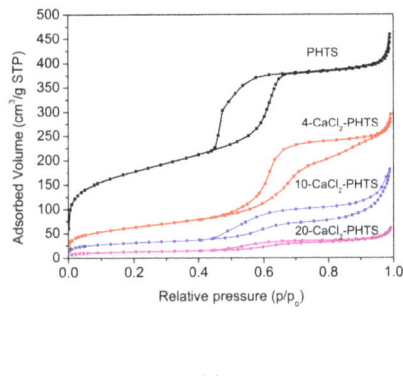 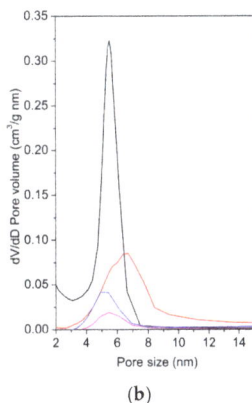

(a) (b)

Figure 6. (a) Nitrogen sorption isotherms and (b) pore size distribution of PHTS and composites after water sorption.

Table 2. Structural properties of PHTS and the composites after water sorption at 40 °C.

Sample After Water Sorption	S_{BET} (m^2/g)	V_{tot} (cm^3/g)	Average Pore Size (nm)
PHTS	640	0.624	5.5
4-CaCl$_2$-PHTS	227	0.394	6.7
10-CaCl$_2$-PHTS	133	0.195	5.2
20-CaCl$_2$-PHTS	50	0.039	5.5

Abbreviations: S_{BET}, the BET surface area; V_{tot}, total pore volume evaluated from adsorption isotherm at the relative pressure 0.96.

XRD patterns of the matrix and the composites after cycling test (Figure 7a) show collapse of the ordered mesostructure into disordered one for the composites containing 10 and 20 wt.% of the salt. Partial collapse is observed for the composite with 4 wt.% of the salt as well. No diffraction peaks of the salt can be observed in Figure 7b presenting high-angle XRD patterns of the composites.

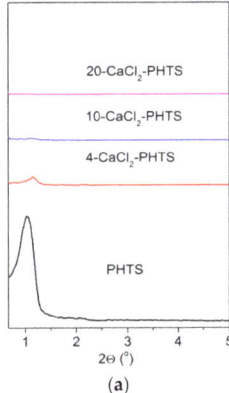 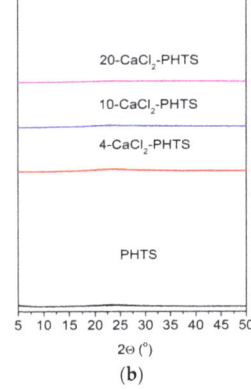

(a) (b)

Figure 7. (a) Low-angle XRD patterns of the matrix and the composites containing different contents of CaCl$_2$ after cycling test and (b) High-angle XRD patterns of the pure matrix and the composites with different amounts of CaCl$_2$ after cycling test.

It can be clearly seen that the shape of the isotherm of the matrix did not change neither after water sorption nor after 20 cycling test (Figure 8). On the other hand, the shape of the nitrogen isotherms of the composites after water sorption and cycling test is significantly different, showing strong influence of water sorption on the structure of the composite's matrix. Porosity (Table 3) of the matrix and the composites were improved, showing the increase of specific surface area, total pore volume, and pore size of the composites. This indicates that salt was still present in the pores after the cycling test, but without salt agglomerates, thus causing further destruction of the mesopores due to corrosiveness of the salt solution and resulting in the increase of pore size of all composites. Interparticle porosity was less pronounced for the PHTS, 4-$CaCl_2$, and 10-$CaCl_2$ samples. Figure 8b shows broad and less intensive pore size distributions of the composites comparing to the matrix after 20 cycles.

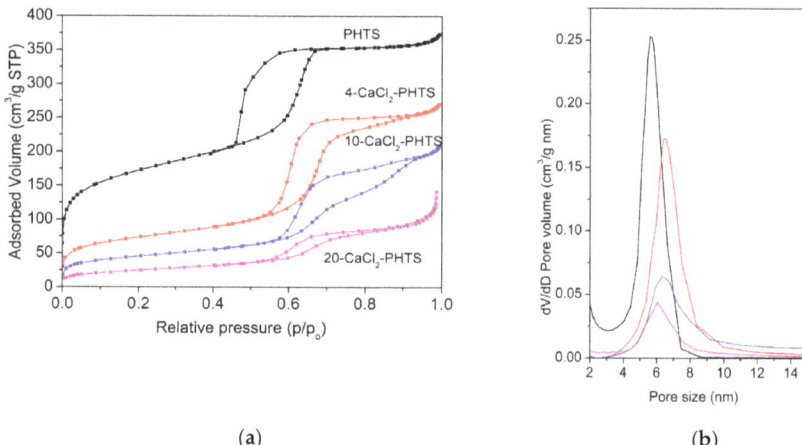

(a)   (b)

Figure 8. (a) Nitrogen sorption isotherms and (b) pore size distribution of PHTS and composites after 20 cycles.

Table 3. Elemental analysis and structural properties of PHTS and the composites after cycling tests between 40 and 140 °C at 56 mbar.

Sample After Cycling	S_{BET} (m²/g)	V_{tot} (cm³/g)	Average Pore Size (nm)	EDX Analysis (wt.%)
PHTS	620	0.560	5.7	-
4-$CaCl_2$-PHTS	256	0.400	6.4	4
10-$CaCl_2$-PHTS	165	0.304	6.5	10
20-$CaCl_2$-PHTS	90	0.119	6.0	20

Abbreviations: S_{BET}, the BET surface area; V_{tot}, total pore volume evaluated from adsorption isotherm at the relative pressure 0.96.

SEM pictures after water sorption are similar to the pictures of the as-prepared samples. The pictures after 20 cycles are presented in Figure 9. It can be seen that morphology of PHTS and all composites did not change after the cycling test.

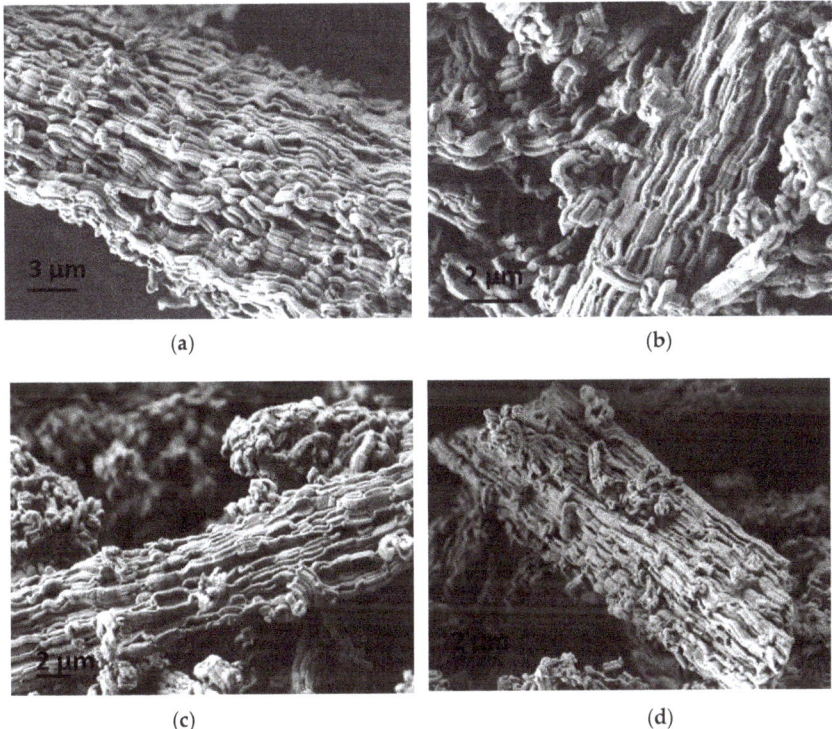

Figure 9. SEM images of (**a**) PHTS matrix; (**b**) 4-CaCl$_2$-PHTS; (**c**) 10-CaCl$_2$-PHTS and (**d**) 20-CaCl$_2$-PHTS after cycling test.

3.3. Water Sorption and Heat Storage Capacity Calculation

Water sorption isotherms performed at 40 °C for the matrix and the composites are shown in Figure 10a. The ordered mesoporous matrix and the composites exhibit sorption of isotherms Type V. The water uptake curve of the matrix showed typical characteristics of weak hydrophilic or hydrophobic mesoporous materials with low sorption at low relative pressure and moderate sorption at the middle relative pressure, and sudden high water sorption at higher relative pressure [22]. The maximal water upload of the matrix was 0.65 g/g, showing the active role [14] of the matrix. A comparison of hydrophilic character of PHTS and SBA-15 from the literature [46] shows that a capillary condensation started at higher p/p$_0$~0.75 for SBA-15 than for PHTS (p/p$_0$~0.65), which indicates that PHTS is more hydrophilic than SBA-15. Another difference of structural property, is evident. Namely, the specific surface area, which influences hydrophilic properties and consequently Qads, of SBA-15 is 554 m^2/g and water uptake of this material is 0.02 g$_{H2O}$/g$_{sample}$ in the 0 < p/p$_0$ < 0.3. On the other hand the specific surface area of PHTS is much higher (810 m^2/g) and the water uptake reaches 0.138 g$_{H2O}$/g$_{sample}$ in the same relative pressure range. It is well known that higher surface area means more available sorption sites in the material and indicates better diffusivity of the vapor, which is crucial for optimal mass and heat transfer. It can be concluded that lower temperature (65 °C) of aging for the PHTS preparation is beneficial for higher water uptake at low relative pressure [23]. Maximal water sorption capacities of the composites increased to 0.78 g/g (4-CaCl$_2$), 1.20 g/g (10-CaCl$_2$), and 2.24 g/g (20-CaCl$_2$). The shape of the uptake curves was evidently changed. For a relative pressure of 0.4, the composite containing 10 wt.% CaCl$_2$, showed double water sorption capacity (0.16 g/g vs. 0.38 g/g), while the composite with 20 wt.% of the salt revealed three times larger water sorption capacity than the matrix (0.16 g/g vs. 0.58 g/g). It can be concluded that the presence of calcium chloride in the matrix increased the water sorption capacity of the composites; thus, the salt content impacted the sorption performance of these composites [12]. On the other hand, the matrix of the composites with the same amount of salt had an important role as well. Namely, comparing water sorption isotherms of 4-CaCl$_2$-PHTS and 4-CaCl$_2$-SBA-15 [26] composites revealed differences in the range $0 < p/p_0 < 0.4$, showing higher uptake for 4-CaCl$_2$-PHTS due to the preparation procedure of PHTS. Water uptake at 0.4 relative pressure of the composite with SBA-15 matrix, possessing uniform mesopores of average pore size of 10.2 nm and larger total pore volume (0.928 cm^3/g), was lower for 0.07 g/g, while the maximum water uptake was higher for 0.10 g/g. The water uptake curve of 20-CaCl$_2$-PHTS showed a plateau at 0.13 p/p$_0$ due to formation of calcium chloride dihydrate [25], while for the composites with lower salt contents this plateau was not observed. The characteristic curves of the matrix and the composites, which showed the adsorbed water uptakes as a function of the adsorption potential A [47,48], are plotted in Figure 10b and are comparable with those previously reported [49].

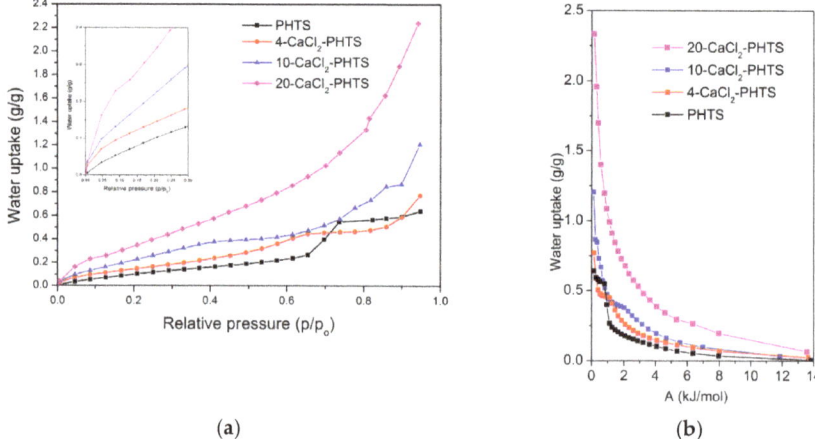

(a) (b)

Figure 10. (a) Water uptake curves for the matrix and the composites (inset: water uptake in the range $0 < p/p_0 < 0.30$) at 40 °C; (b) the characteristic curves for water sorption on the PHTS and the composites.

The most relevant parameter for evaluating the effectiveness of sorbent for TES is the integral heat of adsorption. It is well known that the adsorbents with low water uptake after desorption and high water uptake after adsorption, resulting in high water loading lift to reach high storage densities, are needed for sorption heat storage. The integral heat of adsorption was calculated for the given boundary conditions for space heating [12]: adsorption temperature at 40 °C, desorption temperature at 120 °C, and a dew point temperature was set at 10 °C. The integral heat of adsorption Q_{ads} of all composites is listed in Table 4.

Table 4. Water loading lift and the integral heat of adsorption for the composites.

Sample	Δw (kg/kg)	Q_{ads} (Wh/kg)	Q_{ads} (kJ/kg)
PHTS	0.073	71	256
4-CaCl$_2$-PHTS	0.100	81	292
10-CaCl$_2$-PHTS	0.142	119	428
20-CaCl$_2$-PHTS	0.239	193	694

The increased amount of the salt in the composites increased the calculated water loading lifts, and consequently, the energy storage capacity. A direct comparison of the energy storage capacities of these composites with other composites containing $CaCl_2$ was difficult and risky, because it strongly depends on the boundary conditions. The energy storage capacity value of the $CaCl_2$ (43 wt.%) encapsulated in a silica gel equaled 300 Wh/kg, considering desorption temperatures of 80 °C, adsorption temperature of 30 °C, and adsorption pressure of 12 mbar [12]. The composite of 7 wt.% $CaCl_2$ confined in disordered mesoporous iron silicate matrix shows energy storage capacity of 155 Wh/kg at adsorption temperature of 25 °C, desorption temperature of 150 °C, and adsorption pressure of 12 mbar [13]. The energy storage capacity of 240 Wh/kg can be found for the aluminosilicate containing 30 wt% of the same salt at adsorption temperature of 40 °C, desorption temperature of 120 °C, and adsorption pressure of 20 mbar [15]. Higher values were achieved for MIL-100(Fe)/46 wt.% $CaCl_2$ (335 Wh/kg) and MIL-101(Cr)/62 wt% $CaCl_2$ (485 Wh/kg) at desorption temperature of 80 °C, adsorption temperature of 30 °C, and adsorption pressure of 12 mbar [19].

The influence of desorption and adsorption temperature on the water loading lift was evaluated (Table 5) for the boundary conditions: desorption temperature at 120 °C, adsorption temperature at 30 °C, and due point temperature of 10 °C. It can be seen that lower adsorption temperature led to the increased water loading lifts of the composites, which corresponds to higher energy storage capacity of the composites [12], and thus promotes their applicability for low-temperature thermal energy storage.

Table 5. Water loading lift and the integral heat of adsorption for the composites.

Sample	Δw (kg/kg)	Q_{ads} (Wh/kg)	Q_{ads} (kJ/kg)
PHTS	0.125	117	421
4-$CaCl_2$-PHTS	0.150	131	472
10-$CaCl_2$-PHTS	0.250	205	738
20-$CaCl_2$-PHTS	0.430	333	1199

On the other hand, a lower desorption temperature at 100 °C will decrease the performance of the composites (e.g., theirs energy storage capacities). For example, the calculated water loading lift of the 10-$CaCl_2$-PHTS decreased to 0.100 kg/kg, which corresponds to the energy storage capacity of 86 Wh/kg.

Cycling stability of these composites during 20 cycles of sorption and desorption between 40 and 140 °C at 56 mbar was tested, showing a small reduction of water uptake (2–6%) after the last cycle for each composite. No leaching of the salt from the PHTS matrix was evidenced (Table 3), showing the ability of the PHTS matrix to create a stable nano-environment for confinement of calcium chloride. This confirms that these composites are promising candidates for low-temperature thermal energy storage.

4. Conclusions

Novel composites composed of PHTS (plugged hexagonal templated silicate) with hexagonal pore arrangement as the matrix and 4 wt.%, 10wt.%, and 20 wt.% of calcium chloride have been developed by incipient wetness impregnation. The preparation procedure of the matrix increased its hydrophilic properties, showing its active role for water sorption. Calcium chloride is shown to be located in the pores of the matrix. The presence of $CaCl_2$ in PHTS leads to a partial deformation of ordered pore structure (10-$CaCl_2$) or a collapse of the ordered pore arrangement (20-$CaCl_2$) into the disordered mesostructure. Water sorption caused structural modifications of the composites, showing the re-dispersion and possible agglomeration of the salt in the pores, which caused some blocking of pores (lower total pore volume and specific surface areas) after hydration and dehydration at 40 °C. On the other hand, repeated sorption/desorption cycles between 40 and 140 °C at 56 mbar caused the improvement of structural properties (increase of specific surface area, total pore volume, and pore size) of the 10-$CaCl_2$ and 20-$CaCl_2$ composites, indicating that highly dispersed salt was still present in the pores. The increased salt content in the composites impacted the sorption performance of these composites, e.g., higher content of the salt higher energy storage capacity. An increase of desorption temperature or a decrease of the adsorption temperature increased the water loading lift and consequently the energy storage capacity, as well as the performance of the composites. The comparatively good initial stability of these composites under the operating conditions was determined without any salt leaching. These composites are promising candidates for low-temperature thermal energy storage.

Author Contributions: Conceptualization, A.R.; investigation, A.R.; writing—original draft preparation, A.R.; writing—review and editing, N.Z.L. and A.R.; visualization, A.R.; supervision, A.R.; funding acquisition, N.Z.L.

Funding: Financial support from the Slovenian Research Agency through research program P1-0021 (Nanoporous materials) and research project L1-7665 (Advanced heat storage materials for integrated storage solutions) is acknowledged.

Acknowledgments: We thank Edi Kranjc for XRD measurements, Mojca Opresnik for nitrogen physisorption measurements and EDAX/SEM, Tadeja Birsa Čelič for water sorption measurements, and Matjaž Mazaj for TEM pictures.

Conflicts of Interest: The authors declare no conflict of interest.

References

1. Scapino, L.; Zondag, H.A.; Van Bael, J.; Diriken, J.; Rindt, C.C.M. Sorption heat storage for long-term low-temperature applications: A review on the advancements at material and prototype scale. *Appl. Energy* **2017**, *190*, 920–948. [CrossRef]
2. Mastronardo, E.; Bonaccorsi, L.; Kato, Y.; Piperopoulos, E.; Milone, C. Efficiency improvement of heat storage materials for $MgO/H_2O/Mg(OH)_2$ chemical heat pumps. *Appl. Energy* **2016**, *162*, 31–39. [CrossRef]
3. Ristić, A.; Fischer, F.; Hauer, A.; Zabukovec Logar, N. Improved performance of binder-free zeolite Y for low-temperature sorption heat storage. *J. Mater. Chem. A* **2018**, *6*, 11521–11530. [CrossRef]
4. Grekova, A.; Gordeeva, L.; Aristov, Y. Composite sorbents "Li/Ca halogenides inside Multi-wall Carbon Nano-tubes" for Thermal Energy Storage. *Sol. Energy Mater. Sol. Cells* **2016**, *155*, 176–183. [CrossRef]
5. Henninger, S.K.; Ernst, S.J.; Gordeeva, L.; Bendix, P.; Froechlich, D.; Grekova, A.D.; Bonaccorsi, L.; Aristov, Y.; Jaenchen, J. New materials for adsorption heat transformation and storage. *Renew. Energy* **2017**, *110*, 59–68. [CrossRef]
6. Vasta, S.; Brancato, V.; La Rosa, D.; Palomba, V.; Restuccia, G.; Sapienza, A.; Frazzica, A. Adsorption Heat Storage: State-of-the-Art and Future Perspectives. *Nanomaterials* **2018**, *8*, 522. [CrossRef] [PubMed]
7. Grekova, A.D.; Gordeeva, L.G.; Aristov, Y.I. Composite "LiCl/vermiculite" as advanced water sorbent for thermal energy storage. *Appl. Therm. Eng.* **2017**, *124*, 1401–1408. [CrossRef]
8. Grekova, A.D.; Girnik, I.S.; Nikulin, V.V.; Tokarev, M.M.; Gordeeva, L.G.; Aristov, Y.I. New composite sorbents of water and methanol "salt in anodic alumina": Evaluation for adsorption heat transformation. *Energy* **2016**, *106*, 231–239. [CrossRef]
9. Gordeeva, L.G.; Aristov, Y.I. Composites "salt inside porous matrix" for adsorption heat transformation: A current state-of-the-art and new trends. *Int. J. Low-Carbon Technol.* **2012**, *7*, 288–302. [CrossRef]
10. Gaeini, M.; Rouws, A.L.; Salari, J.W.O.; Zondag, H.A.; Rindt, C.C.M. Characterization of microencapsulated and impregnated porous host materials based on calcium chloride for thermochemical energy storage. *Appl. Energy* **2018**, *212*, 1165–1177. [CrossRef]
11. Aristov, Y.I. New family of solid sorbents for adsorptive cooling: Material scientist approach. *J. Eng. Thermophys.* **2007**, *16*, 63–72. [CrossRef]
12. Courbon, E.; D'Ans, P.; Permyakova, A.; Skrylnyk, O.; Steunou, N.; Degrez, M.; Frère, M. Further improvement of the synthesis of silica gel and $CaCl_2$ composites: Enhancement of energy storage density and stability over cycles for solar heat storage coupled with space heating applications. *Sol. Energy* **2017**, *157*, 532–541. [CrossRef]
13. Ristić, A.; Maučec, D.; Henninger, S.K.; Kaučič, V. New two-component water sorbent $CaCl_2$-FeKIL2 for solar thermal energy storage. *Microporous Mesoporous Mater.* **2012**, *164*, 266–272. [CrossRef]

14. Ponomarenko, I.V.; Glaznev, I.S.; Gubar, A.V.; Aristov, Y.I.; Kirik, S.D. Synthesis and water sorption properties of a new composite "CaCl$_2$ confined into SBA-15 pores". *Microporous Mesoporous Mater.* **2010**, *129*, 243–250. [CrossRef]
15. Jänchen, J.; Ackermann, D.; Stach, H.; Brösicke, W. Studies of the water adsorption on Zeolites and modified mesoporous materials for seasonal storage of solar heat. *Sol. Energy* **2004**, *76*, 339–344. [CrossRef]
16. Nonnen, T.; Beckert, S.; Gleichmann, K.; Brandt, A.; Unger, B.; Kerskes, H.; Mette, B.; Bonk, S.; Badenhop, T.; Salg, F.; et al. A Thermochemical Long-Term Heat Storage System Based on a Salt/Zeolite Composite. *Chem. Eng. Technol.* **2016**, *39*, 2427–2434. [CrossRef]
17. Jabbari-Hichri, A.; Bennici, S.; Auroux, A. CaCl$_2$-containing composites as thermochemical heat storage materials. *Sol. Energy Mater. Sol. Cells* **2017**, *172*, 177–185. [CrossRef]
18. Casey, S.P.; Elvins, J.; Riffat, S.; Robinson, A. Salt impregnated desiccant matrices for "open" thermochemical energy storage—Selection, synthesis and characterisation of candidate materials. *Energy Build.* **2014**, *84*, 412–425. [CrossRef]
19. Permyakova, A.; Wang, S.; Courbon, E.; Nouar, F.; Heymans, N.; D'Ans, P.; Barrier, N.; Billemont, P.; De Weireld, G.; Steunou, N.; et al. Design of salt-metal organic framework composites for seasonal heat storage applications. *J. Mater. Chem. A* **2017**, *5*, 12889–12898. [CrossRef]
20. Palomba, V.; Frazzica, A. Recent advancements in sorption technology for solar thermal energy storage applications. *Sol. Energy* **2018**, 1–37. [CrossRef]
21. N'Tsoukpoe, K.E.; Rammelberg, H.U.; Lele, A.F.; Korhammer, K.; Watts, B.A.; Schmidt, T.; Ruck, W.K.L. A review on the use of calcium chloride in applied thermal engineering. *Appl. Therm. Eng.* **2015**, *75*, 513–531. [CrossRef]
22. Ng, E.P.; Mintova, S. Nanoporous materials with enhanced hydrophilicity and high water sorption capacity. *Microporous Mesoporous Mater.* **2008**, *114*, 1–26. [CrossRef]
23. Maaz, S.; Rose, M.; Palkovits, R. Systematic investigation of the pore structure and surface properties of SBA-15 by water vapor physisorption. *Microporous Mesoporous Mater.* **2016**, *220*, 183–187. [CrossRef]
24. Tokarev, M.; Gordeeva, L.; Romannikov, V.; Glaznev, I.; Aristov, Y. New composite sorbent CaCl$_2$ in mesopores for sorption cooling/heating. *Int. J. Therm. Sci.* **2002**, *41*, 470–474. [CrossRef]
25. Glaznev, I.; Ponomarenko, I.; Kirik, S.; Aristov, Y. Composites CaCl$_2$/SBA-15 for adsorptive transformation of low temperature heat: Pore size effect. *Int. J. Refrig.* **2011**, *34*, 1244–1250. [CrossRef]
26. Ristić, A.; Henninger, S.K. Sorption composite materials for solar thermal energy storage. *Energy Procedia* **2014**, *48*, 977–981. [CrossRef]
27. Aristov, Y.I. Challenging offers of material science for adsorption heat transformation: A review. *Appl. Therm. Eng.* **2013**, *50*, 1610–1618. [CrossRef]
28. Thommes, M.; Kaneko, K.; Neimark, A.V.; Olivier, J.P.; Rodriguez-Reinoso, F.; Rouquerol, J.; Sing, K.S.W. Physisorption of gases, with special reference to the evaluation of surface area and pore size distribution (IUPAC Technical Report). *Pure Appl. Chem.* **2015**, *87*, 1051–1069. [CrossRef]
29. Van Der Voort, P.; Ravikovitch, P.I.; De Jong, K.P.; Benjelloun, M.; Van Bavel, E.; Janssen, A.H.; Neimark, A.V.; Weckhuysen, B.M.; Vansant, E.F. A new templated ordered structure with combined micro- and mesopores and internal silica nanocapsules. *J. Phys. Chem. B* **2002**, *106*, 5873–5877. [CrossRef]
30. Celer, E.B.; Kruk, M.; Zuzek, Y.; Jaroniec, M. Hydrothermal stability of SBA-15 and related ordered mesoporous silicas with plugged pores. *J. Mater. Chem.* **2006**, *16*, 2824. [CrossRef]
31. Zhao, D.; Huo, Q.; Feng, J.; Chmelka, B.F.; Stucky, G.D. Nonionic Triblock and Star Diblock Copolymer and Oligomeric Surfactant Syntheses of Highly Ordered, Hydrothermally Stable, Mesoporous Silica Structures. *J. Am. Chem. Soc.* **1998**, *120*, 6024–6036. [CrossRef]
32. Munnik, P.; De Jongh, P.E.; De Jong, K.P. Recent Developments in the Synthesis of Supported Catalysts. *Chem. Rev.* **2015**, *115*, 6687–6718. [CrossRef] [PubMed]
33. Brunauer, S.; Emmett, P.H.; Teller, E. Adsorption of Gases in Multimolecular Layers. *J. Am. Chem. Soc.* **1938**, *60*, 309–319. [CrossRef]
34. Barrett, E.P.; Joyner, L.G.; Halenda, P.P. The Determination of Pore Volume and Area Distributions in Porous Substances. I. Computations from Nitrogen Isotherms. *J. Am. Chem. Soc.* **1951**, *73*, 373–380. [CrossRef]
35. de Lange, M.F.; Verouden, K.J.F.M.; Vlugt, T.J.H.; Gascon, J.; Kapteijn, F. Adsorption-Driven Heat Pumps: The Potential of Metal–Organic Frameworks. *Chem. Rev.* **2015**, *115*, 12205–12250. [CrossRef] [PubMed]

36. Bering, B.P.; Dubinin, M.M.; Serpinsky, V.V. Theory of volume filling for vapor adsorption. *J. Colloid Interface Sci.* **1966**, *21*, 378–393. [CrossRef]
37. Krajnc, A.; Varlec, J.; Mazaj, M.; Ristić, A.; Logar, N.Z.; Mali, G. Superior Performance of Microporous Aluminophosphate with LTA Topology in Solar-Energy Storage and Heat Reallocation. *Adv. Energy Mater.* **2017**, *7*, 1601815. [CrossRef]
38. Mazaj, M.; Stevens, W.J.J.; Logar, N.Z.; Ristić, A.; Tušar, N.N.; Arčon, I.; Daneu, N.; Meynen, V.; Cool, P.; Vansant, E.F.; et al. Synthesis and structural investigations on aluminium-free Ti-Beta/SBA-15 composite. *Microporous Mesoporous Mater.* **2009**, *117*, 458–465. [CrossRef]
39. Tasbihi, M.; Štangar, U.L.; Škapin, A.S.; Ristić, A.; Kaučič, V.; Tušar, N.N. Titania-containing mesoporous silica powders: Structural properties and photocatalytic activity towards isopropanol degradation. *J. Photochem. Photobiol. A Chem.* **2010**, *216*, 167–178. [CrossRef]
40. Šuligoj, A.; Štangar, U.L.; Ristić, A.; Mazaj, M.; Verhovšek, D.; Tušar, N.N. TiO2-SiO2films from organic-free colloidal TiO2anatase nanoparticles as photocatalyst for removal of volatile organic compounds from indoor air. *Appl. Catal. B Environ.* **2016**, *184*, 119–131. [CrossRef]
41. Lemaire, A.; Rooke, J.C.; Chen, L.H.; Su, B.L. Direct observation of macrostructure formation of hierarchically structured meso-macroporous aluminosilicates with 3D interconnectivity by optical microscope. *Langmuir* **2011**, *27*, 3030–3043. [CrossRef] [PubMed]
42. Lukens, W.W.; Schmidt-Winkel, P.; Zhao, D.; Feng, J.; Stucky, G.D. Evaluating pore sizes in mesoporous materials: A simplified standard adsorption method and a simplified Broekhoff-de Boer method. *Langmuir* **1999**, *15*, 5403–5409. [CrossRef]
43. Tanev, P.T.; Vlaev, L.T. An attempt at a more precise evaluation of the approach to mesopore size distribution calculations depending on the degree of pore blocking. *J. Colloid Interface Sci.* **1993**, *160*, 110–116. [CrossRef]
44. Deshmane, V.G.; Adewuyi, Y.G. Mesoporous nanocrystalline sulfated zirconia synthesis and its application for FFA esterification in oils. *Appl. Catal. A Gen.* **2013**, *462*, 196–206. [CrossRef]
45. Pikus, S.; Celer, E.B.; Jaroniec, M.; Solovyov, L.A.; Kozak, M. Studies of intrawall porosity in the hexagonally ordered mesostructures of SBA-15 by small angle X-ray scattering and nitrogen adsorption. *Appl. Surf. Sci.* **2010**, *256*, 5311–5315. [CrossRef]
46. Jabbari-Hichri, A.; Bennici, S.; Auroux, A. Effect of aluminum sulfate addition on the thermal storage performance of mesoporous SBA-15 and MCM-41 materials. *Sol. Energy Mater. Sol. Cells* **2016**, *149*, 232–241. [CrossRef]
47. Dubinin, M.M. The potential theory of adsorption of gases and vapors for adsorbents with energetically nonuniform surfaces. *Chem. Rev.* **1960**, *60*, 235–241. [CrossRef]
48. Birsa Čelič, T.; Mazaj, M.; Guillou, N.; Elkaïm, E.; El Roz, M.; Thibault-Starzyk, F.; Mali, G.; Rangus, M.; Čendak, T.; Kaučič, V.; et al. Study of hydrothermal stability and water sorption characteristics of 3-dimensional Zn-based trimesate. *J. Phys. Chem. C* **2013**, *117*, 14608–14617. [CrossRef]
49. Aristov, Y. Concept of adsorbent optimal for adsorptive cooling/heating. *Appl. Therm. Eng.* **2014**, *72*, 166–175. [CrossRef]

© 2018 by the authors. Licensee MDPI, Basel, Switzerland. This article is an open access article distributed under the terms and conditions of the Creative Commons Attribution (CC BY) license (http://creativecommons.org/licenses/by/4.0/).

Article

Hydrated Salt/Graphite/Polyelectrolyte Organic-Inorganic Hybrids for Efficient Thermochemical Storage

Sergio Salviati [1,2], Federico Carosio [1,*], Guido Saracco [1,2] and Alberto Fina [1]

1. Dipartimento di Scienza Applicata e Tecnologia, Politecnico di Torino-Alessandria Campus, 15121 Alessandria, Italy; sergio.salviati@polito.it (S.S.); guido.saracco@polito.it (G.S.); alberto.fina@polito.it (A.F.)
2. Center for Sustainable Future Technologies, Istituto Italiano di Tecnologia, 10144 Torino, Italy
* Correspondence: federico.carosio@polito.it

Received: 22 January 2019; Accepted: 6 March 2019; Published: 12 March 2019

Abstract: Hydrated salt thermochemical energy storage (TES) is a promising technology for high density energy storage, in principle opening the way for applications in seasonal storage. However, severe limitations are affecting large scale applications, related to their poor thermal and mechanical stability on hydration/dehydration cycling. In this paper, we report the preparation and characterization of composite materials manufactured with a wet impregnation method using strontium bromide hexahydrate (SBH) as a thermochemical storage material, combined with expanded natural graphite (G). In addition to these fully inorganic formulations, an organic polyelectrolyte (PDAC, polydiallyldimethylammonium chloride) was exploited in the structure, with the aim to stabilize the salt, while contributing to the sorption/desorption process. Different formulations were prepared with varying PDAC concentration to study its contribution to material morphology, by electron microscopy and X-ray diffraction, as well as water sorption/desorption properties, by thermogravimetry and differential calorimetry. Furthermore, the SBH/G/PDAC powder mixture was pressed to form tabs that were analyzed in a climatic chamber, which is evidence for an active role of PDAC in the improvement of water sorption, coupled with a significant enhancement of mechanical resistance upon hydration/dehydration cycling. Therefore, the addition of the polyelectrolyte is proposed as an innovative approach in the fabrication of efficient and durable TES devices.

Keywords: thermal energy storage; thermochemical energy storage; salt hydration; composite materials; heat transfer; mass transfer

1. Introduction

Fighting climate change is one of the biggest challenges, attracting research efforts from all over the world. Smart heat management is presently a central topic in greenhouse gas mitigation and the approach of thermal energy storage (TES) has a key role in achieving this goal [1]. The simple and fundamental concept at the base of TES is to identify heat sources with low environmental and economic impact, store their energy when it is not needed, and use it in later times instead of producing it with other traditional alternatives. These technologies have been proven to have a potential reduction in CO_2 emissions up to 5.5% compared to 1990 levels [2]. In particular, low temperature heat sources (up to 150 °C) are considered among the greatest opportunities in this field. These sources can be identified mainly in two scenarios: Waste heat reuse and smart heat management in buildings. In the first case, low grade heat that is commonly released to the environment may in principle be conveniently recovered and stored for later use. Several sources were identified in different industrial fields such as oil, chemicals, steel, glass [3], and food [4] industries, as well as municipal solid waste, mining wastes [5], data centers [6], and car engines [7]. On the other hand, one of the main issues in energy management in the building sector is the mismatch between heat supply and demand. In this field, two main areas are identified: Short-term storage (e.g., from night to day), and long-term storage (e.g., from summer to winter). Following these needs, many TES devices were successfully developed and implemented in both residential and commercial structures [8]. In addition to this, the increasing development of renewable energy sources creates the need to reinvent energy demand management resulting in integrated solutions where TES technologies are coupled with photovoltaic panels and/or solar thermal technologies in order to manage the peak of electricity demand and reduce the costs related to electricity consumption [9,10].

The most used classification of TES materials takes into account the form in which heat is stored. Sensible heat storage is the most widely adopted and well-known techniques, because it is based on cheap and highly available materials (e.g., water or concrete) with high specific heat. The second approach uses the latent heat of phase change materials (PCMs), such as paraffins, to obtain higher energy storage densities with respect to sensible heat [11]. The third class is thermochemical TES and it includes materials showing a high-enthalpy reversible gas/solid reaction. One of the most promising thermochemical materials (TCMs) class is inorganic salt hydrates ($M_nA_m \cdot XH_2O$) [12], in which the storage reaction is:

$$M_nA_m \cdot XH_2O + heat \leftrightarrow M_nA_m \cdot (X-Y)H_2O + YH_2O$$

When heat is transferred from a selected source to the TCM, in the so-defined charging step, dehydration occurs. As long as the salt is maintained in the dehydrated state, latent heat is stored. When water is made available to the salt (discharging step), hydration occurs and hydration heat is released. The possibility to control the heat release by controlling the water feed to the dehydrated salt, is one of the main advantages of this technique, making the heat discharge controllable on demand [13]. The second important advantage of TCMs over PCMs is the higher (one order of magnitude) energy storage density associated with the employed materials [14]. Due to this great potential, many efforts were made to identify the best performing salt hydrates with both experimental [15,16], and theoretical methods [17,18]. Despite the selection of the hydrated salt primarily depending on the available source temperature, one of the most promising TCMs discussed in literature for low temperature TES applications is $SrBr_2 \cdot 6H_2O$ (SBH) owing to its effective combination of a relatively high storage density (798 kJ/kg) and low dehydration temperature ($\simeq 100\ ^\circ C$) as reviewed in a detailed study [19]. Unfortunately, first efforts to implement this salt in a working device also showed some severe limitations, in terms of low thermal conductivity, low chemical stability over hydration/dehydration cycles, and slow mass/heat transfer [20]. Research efforts were mostly aimed at overcoming these limits by design improvement, while few studies dealt with new material concepts in order to fully exploit the potentialities of SBH. One practical approach is to include the TCMs in a porous matrix to overcome the drawbacks of solid/gas reactions [21]. In particular, expanded natural graphite (G) was proposed for combination with TCMs, based on its low price and density, coupled with high thermal conductivity and surface area. Recently, hybrid salt/graphite materials were prepared and tested [22–24], but in most cases an inherent incompatibility between the structures of the two materials resulted in big salt aggregates, thus minimizing water adsorption kinetics, and the rate of heat transfer between salt and graphite layers. In this manuscript, we aim at overcoming these limitations by producing graphite composites encompassing a polyelectrolyte binder, PDAC (polydiallyldimethylammonium chloride), to enhance the compatibility between salt and matrix. Indeed, PDAC is known to have a strong interaction with graphite layers [25], and it is expected to show a good affinity with ionic materials due to its high charge density. The aim is to obtain a better distribution of salt on the pores surfaces, maximizing the area of the air/salt and salt/graphite interfaces. In addition, PDAC also shows good moisture sorption ability, thus potentially improving the hydration kinetics of the salt [26].

2. Experimental Section

2.1. Materials

PDAC (Mw = 400,000–500,000 g/mol) was purchased from Sigma-Aldrich® as 20% wt/wt water solution. In order to obtain solid PDAC samples for the analyses, the solution was dried in an oven at 120 °C until constant weight was reached and hydrated in a climatic chamber at 23 °C and 50% relative humidity (RH) overnight. Expanded natural graphite (G), with 28.4 m^2/g surface area (as reported in the material datasheet) was purchased by TIMCAL (Bodio, Switzerland), commercial grade TIMREX® BNB90. SrBr$_2$·6H$_2$O (S) with >95% purity in powder form was purchased from Alfa Aesar® (Haverhill, MA, US). All reagents were used as received for preparing stable water dispersions using deionized water supplied by a Direct-Q® 3 UV Millipore System (Milano, Italy).

2.2. TCM Composite Manufacturing

The main steps in the manufacturing process are depicted in Figure 1.

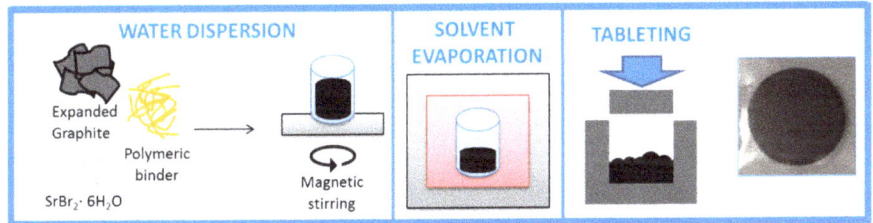

Figure 1. Thermochemical material (TCM) composite material manufacturing process.

PDAC water solution was diluted with 30 mL of water with subsequent additions of G and SBH. Four samples were prepared varying the amount of polymer while keeping the G/SBH ratio constant (their compositions are shown in Table 1).

Table 1. Summary of the prepared samples.

Sample Name	Weight Ratio		
	G	SBH	PDAC
	Expanded Natural Graphite	Strontium Bromide Hexahydrate	Polyelectrolyte Binder
SBH/G	1	5	0
SBH/G/P 0.1	1	5	0.1
SBH/G/P(0.5)	1	5	0.5
SBH/G/P(1)	1	5	1

The suspension was stirred overnight to obtain a homogeneous dispersion and then heated at 100 °C on a plate while stirring for around 5 h to remove water via evaporation. In this step, SBH and PDAC started precipitating on the graphite matrix. When the dispersion viscosity was too high to allow any further stirring, the wet material was placed in a vacuum oven at 50 °C overnight to complete the process. After these steps, the prepared mixtures were subjected to a complete dehydration and hydration cycle prior to the subsequent charge/discharge cycles [27]. The samples were dehydrated in a ventilated oven at 120 °C until they reached constant weight and rehydrated in a climatic chamber at 23 °C and 50% RH. After that the samples were tableted using a stainless steel mold in a hydraulic press under the pressure of 1 t. The nominal tab size was 30 mm in diameter and 3 mm in height.

2.3. Characterization

The materials morphology was investigated with a LEO-1450VP (Zeiss, Oberkochen, Germany) scanning electron microscope (SEM) with a 15 kV accelerating voltage, on the cross section of tabs, obtained by fragile fracture upon bending. Surfaces were gold-coated prior to SEM observations.

XRD analyses were performed on a Philips/Panalytical X´Pert Pro (Malvern, Milano, Italy) using a Philips PW3040/60 X-ray generator with a Cu anode using a Kα wavelength. A broad interval of 2θ angles of 10–70 were chosen to identify the SBH structure using a 0.026° 2θ as scan step and nominal time per step of 100 s, using a scanning PixCell 1d detector. Intensity of reported diffractograms was normalized on an SBH (110) peak.

The performance of the composite materials was investigated with both differential scanning calorimetry (DSC) on a TA Instruments Q20 system (TA Instruments, Milano, Italy) using open aluminum pans and thermogravimetric analysis (TGA) on a TA Instruments Discovery gravimetric balance using open platinum pans. Both experiments were performed with the same temperature program: An equilibration at 35 °C, a heating ramp to 90 °C at 10 °C/min, and an isotherm for 90 min with a dry nitrogen flux of 50 mL/min for DSC and 25 mL/min for TGA. Only for the PDAC sample, was the isotherm time set to 10 h to assure complete dehydration. Samples weight was ≈7 ± 0.5 mg.

Thermal conductivity tests of the prepared tabs were carried out on a TPS 2500S by Hot Disk AB (Göteborg, Sweden) with a Kapton sensor (radius 6.4 mm) using the slab method [28]. Before each measurement, specimens were stored in a constant climate chamber (Binder KBF 240, Tuttlingen, Germany) at 23.0 ± 0.1 °C and 50.0 ± 0.1% RH for at least 48 h before tests. The test temperature (23.00 ± 0.01 °C) was controlled by a silicon oil bath (Haake A40, Thermo Scientific Inc., Waltham, MA USA) equipped with a temperature controller (Haake AC200, Thermo Scientific Inc., Waltham, MA, USA).

A custom setup was assembled to observe the hydration of the prepared composites. The tabs were vertically held to maximize the surface area exposed to the environment. They were dehydrated in an oven at 120 °C until they reached constant weight. After that they were placed in a climatic chamber (Binder KBF 240, D) at 23.0 ± 0.1 °C, and 50.0 ± 0.1% RH, and weighted on an analytical balance (Radwag AS 220.R2, PL) with an accuracy of ±0.5 mg to record the hydration over time. An experimental deviation of ±10% on the normalized mass gain during rehydration was estimated, after having performed several tests.

3. Results and Discussion

3.1. Morphology Analysis

As the microstructure of the TCM may have affected the kinetic of hydration/dehydration, SEM was employed to investigate the influence of PDAC concentration on the morphology and microstructure of the prepared samples.

The micrographs (Figure 2) unveil the effect of PDAC in the salt distribution. In particular, in the absence of PDAC, salt aggregates in globular shapes with dimensions in the order of few μm were observed between flakes of expanded graphite. In the presence of PDAC, the shape of salt aggregation changes evolved with polyelectrolyte concentration. Indeed, the average size of globular salt agglomerates were reduced in the sample with a PDAC/G ratio of 0.1 (PDAC content of 2% w/w, Figure 2b), and completely absent in the samples with ratio 0.5 (PDAC content of 10% w/w of PDAC) and 1 (PDAC content of 18% w/w of PDAC), as shown in Figure 2c,d respectively. It appears that the PDAC acted as a binder between salt crystals as well as an adhesion promoter at the salt/graphite interface, as schematized in Figure 2e. XRD analysis was used to identify the crystal structure of the employed salt, in particular, it was used to investigate possible anion exchange reactions between SBH and the polyelectrolyte during the water dissolution and recrystallization process. Diffraction patterns for SBH/G and the counterparts with different PDAC concentrations are reported in of Figure 3, while the collected diffractograms for purchased SBH and G are reported in Figure S1.

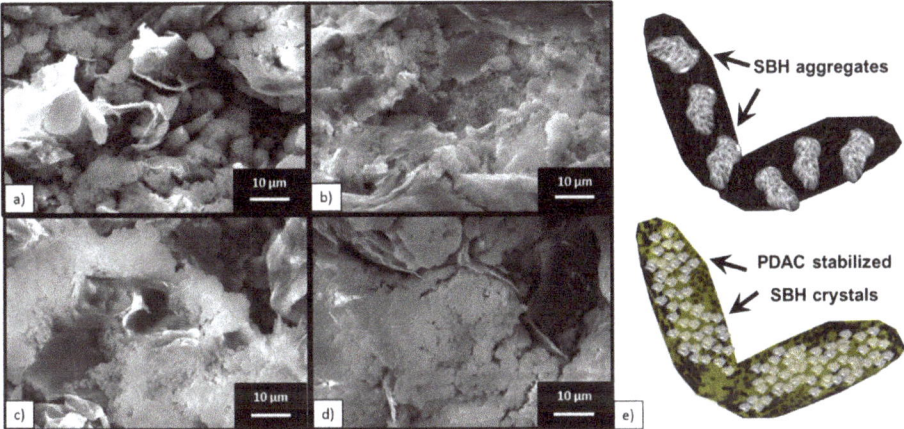

Figure 2. SEM representative images of samples cross sections. (**a**) Strontium bromide hexahydrate/expanded natural graphite (SBH/G), (**b**) SBH/G/P(0.1), (**c**) SBH/G/P(0.5) (**d**) SBH/G/P(1). (**e**) Illustration of the polydiallyldimethylammonium chloride (PDAC) effect in the SBH/G mixture.

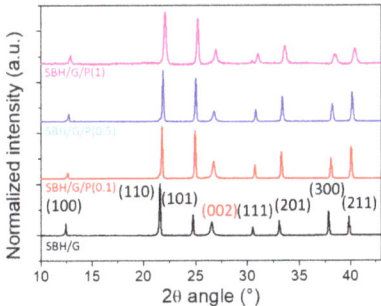

Figure 3. XRD diffractograms of the prepared samples. Miller indices are depicted in black for SBH and in red for graphite crystal planes. Intensity is normalized on the SBH (110) peak.

Using the Joint Committee on Powder Diffraction Standards-International Centre for Diffraction Data database (JCPDS-ICDD) [29], both $SrBr_2 \cdot 6H_2O$ and graphite crystal planes were identified, their Miller indices being reported on the diffractogram in black and red, respectively. The absence of additional peaks excluded the formation of crystalline byproducts derived from ion exchanges between PDAC and SBH during the manufacturing process. Nonetheless, differences in relative intensities of selected peaks (i.e., signals at 24.7° and 39.8°) were observable in the diffractogram between samples with and without polyelectrolyte addition. This could be ascribed to the influence of the polyelectrolyte in the growth of salt hydrates crystals, as previously reported in the literature [30]. In addition, a limited broadening of the SBH main peaks was observed at the highest PDAC concentration, confirming the binder role in the aggregation of SBH crystals. Finally, the absence of extra peaks in PDAC-containing composites confirmed the amorphous nature of the polyelectrolyte.

3.2. Thermal Properties

DSC and TGA analyses were first employed to study the dehydration of the composites in dry conditions. The temperature program was chosen to simulate a 90 °C heat source charging the thermochemical system. DSC results are reported in Figure 4, showing heat flow plots for different samples, characterized by an endothermic peak corresponding to the dehydration reaction.

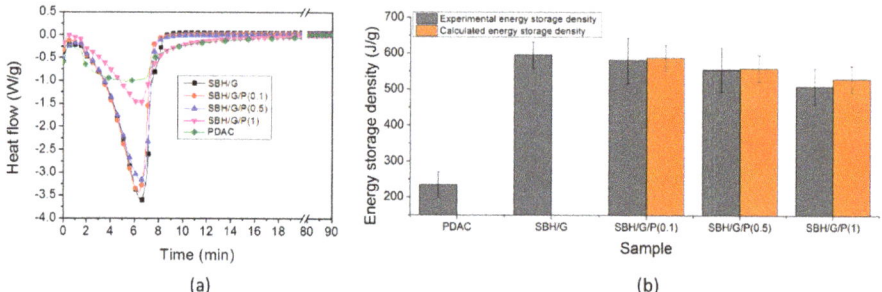

Figure 4. (a) Differential scanning calorimetry (DSC) curves of the prepared samples. (b) Experimental and calculated values for energy storage density.

Figure 4b reports the energy storage density values, calculated from the integral of heat flow plots (average of at least three measurements). For SBH/G/PDAC mixtures, the experimental values were compared with the theoretical values, calculated according with the rule of mixture (Equation (1)), based on the individual components of energy storage density, and their concentration in the mixture.

$$E_c = x_p E_p + x_s E_s \qquad (1)$$

where E_c is the energy density of the composite materials with SBH, G, and PDAC, E_p is the energy density of PDAC, E_s is the energy density of SBH/G and x_p and x_s are the weight fractions of PDAC and SBH/G in the final composites, respectively. As reported in Figure 4b, the experimental values were corresponding to the expected ones, within experimental deviations; this points out that the system acts as an ideal mixture of the two components, with no synergic nor antagonist interaction between PDAC and SBH, in terms of total energy stored. By increasing the content of PDAC, a lowering of the total energy storage density was obtained, diminishing the efficiency of the composite for thermal storage applications by approximately 15% at the highest PDAC concentration. This was ascribed to the great difference in energy storage density between PDAC and the SBH, which was related to the different hydration mechanisms of the two substances. The dehydration of prepared samples were also evaluated by TGA measurements allowing for the assessment of the amount and kinetics of water removal upon heating, as a function of the polyelectrolyte concentration (Figure 5).

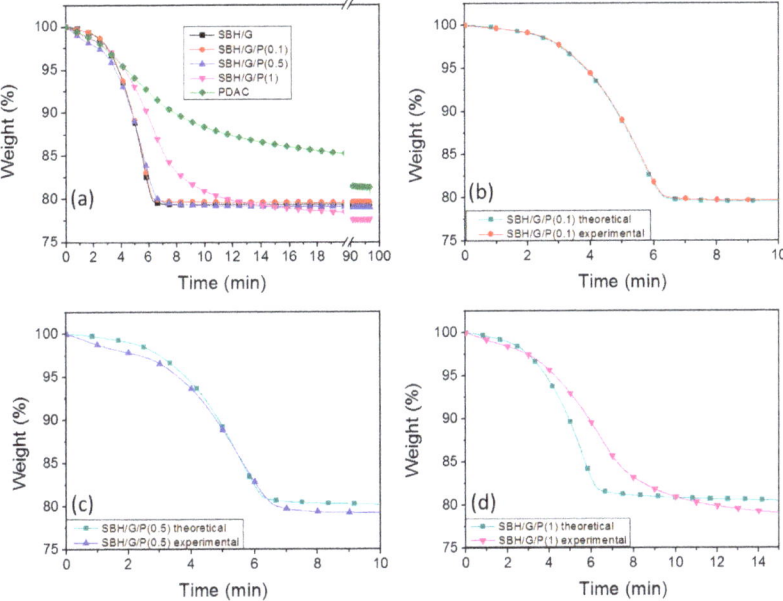

Figure 5. Thermogravimetric analysis (TGA) weight plots of the different SBH/G/PDAC mixtures, compared with pristine PDAC (**a**), and comparison between experimental end calculated data for SBH/G/P(0.1) (**b**), SBH/G/P(0.5) (**c**), and SBH/G/P(1) (**d**).

At low PDAC concentrations (0.1 and 0.5 PDAC/G weight ratios), the polyelectrolyte did not significantly alter the dehydration kinetics with respect to the SBH/G composite. On the other hand, SBH/G/P(1) exhibited a delayed weight loss compared to SBH/G, reflecting the slow dehydration kinetic observed for PDAC. To further investigate the effect of PDAC, the theoretical weight loss curves (W_{th}) for the composites were calculated by applying a rule of mixture between the neat PDAC and SBH/G (Equation (2)).

$$W_c(t) = x_p W_p(t) + x_s W_s(t) \qquad (2)$$

where W_p and x_p are the weight and mass fraction of PDAC in the composite, while W_s and x_s are the weight and mass fraction of SBH in the composite, respectively. As reported in Figure 5b,c, the samples with 0.1 or 0.5 weight ratios showed limited differences between the theoretical and experimental plots, thus suggesting two dehydration processes, from PDAC and SBH, to proceed independently. For SBH/G/P(1), a significant deviation was observed between theoretical and experimental plots, suggesting that kinetics of dehydration were controlled by the interaction between the two phases. This is consistent with the polyelectrolyte binding action between the salt crystals, observed by SEM and ascribed to the delayed diffusion of the water, released by the salt, through the polyelectrolyte. Indeed, while the release of water from crystalline hydrated SBH was simply triggered by the temperature, the amorphous structure of PDAC, with its high free volume, broadened the water release in time, through a series of absorption/desorption steps, eventually reduced the overall dehydration rate. The above results suggest that a high concentration of polyelectrolyte binder can partially reduce the efficiency of the thermochemical system under study by both decreasing the heat storage density (Figure 4b) and slowing down water dehydration kinetics (Figure 5d). Thus, only composites with a PDAC/G weight ratio of 0.1 and 0.5 were selected for tablet preparation and hydration kinetics characterization.

3.3. Composite Tabs Hydration

It is known that the surface/volume ratio can strongly influence the heat and mass transfer phenomena in a thermochemical storage system [23]. For this reason, while values collected with dry TGA and DSC analyses may be used to compare the performance of different composites, these are not representative of a real application, both in terms of mass and moisture effects. In order to obtain more realistic results, hydration kinetics have been evaluated in a climatic chamber on composites tabs having dimensions suitable for real applications. Hydration curves, calculated as the measured weight gain normalized over the weight of the dry tabs, are reported in Figure 6a.

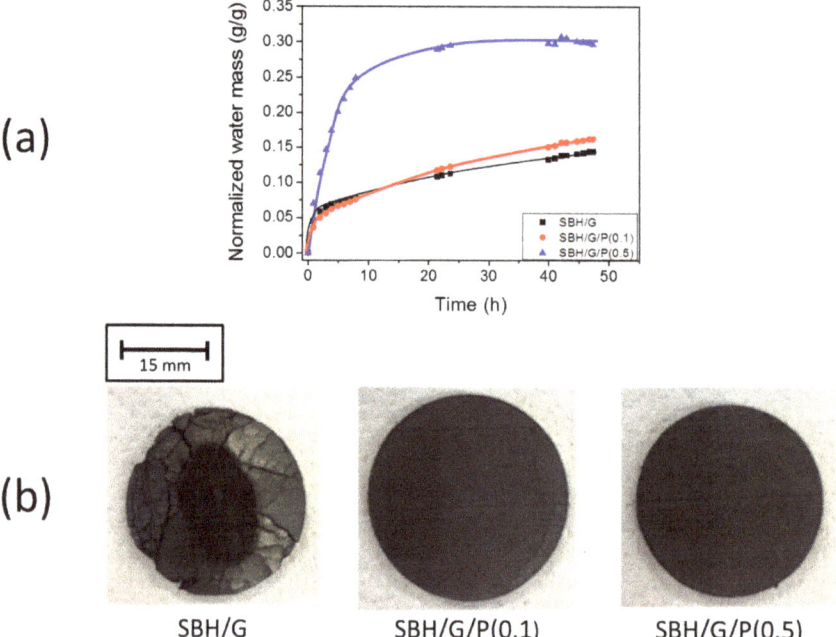

Figure 6. (a) Hydration test on composites tabs; (b) pictures of the tabs after one dehydration/hydration cycle.

The plots clearly show a monotonic weight gain for both SBH/G and counterparts including PDAC. However, dramatic differences in moisture absorption kinetics were proven as a function of PDAC concentration. Indeed, while limited differences existed between SBH/G/P(0.1) and SBH/G, the hydration rate of SBH/G/P(0.5) was much higher, especially within the first hours of the test, reaching the full hydration of both phases in the mixture (equivalent to 0.26 $g_{water}/g_{mixture}$ calculated on the basis of Equation (2)) within approximately 10 h, whereas hydration of the other samples was still ongoing after 45 h (Figure 6a). Another important aspect to evaluate when considering a real application was represented by the durability of the prepared composite. This had a potentially strong impact on the effectiveness and practicability of the thermochemical storage solution. In the present study we observed severe damage in the form of cracks to the SBH/G tabs immediately after the first hydration/dehydration cycle (Figure 6b), thus proving this aspect to be a significant weakness of the graphite/salt hydrate composite approach. The mechanical stress in the samples may have been caused by the volume change of the salt during the hydration process. In fact, it is reported that the density of strontium bromide changed between 3.5 g/cm^3 and 2.4 g/cm^3 from hexahydrate to monohydrate form [31]. On the other hand, the presence of a polymeric binder appeared to reduce crack formation at a polymer/G ratio of 0.1, and completely prevent it a 0.5 ratio, thus maintaining the structural integrity of the composites.

In addition, the thermal conductivity of prepared composite tabs was also evaluated, as heat exchange was obviously crucial for the efficiency of heat storage devices. Results reported in Figure S2 show that the thermal conductivity of the prepared composites remained constant within the experimental error, in the 16–16.5 W/mK range, demonstrating no detrimental effects related to the presence of the polyelectrolyte. Furthermore, thermal conductivity values obtained in this work were significantly higher than previously reported values for similar graphite SBH composites. [24,32]. As clearly depicted by the characterization reported in Figure 6, the composite with a PDAC/G ratio of 0.5 was capable of achieving superior water adsorption kinetics while maintaining high thermal conductivity values thus proving that the inclusion of a polymer binder was a successful strategy for the design of an efficient thermochemical storage solution.

4. Conclusions

This work was focused on the production of composites comprising of strontium bromide hexahydrate, expanded natural graphite, and polydiallyldimethylammonium chloride for thermochemical energy storage applications, using a simple and environmentally sustainable, water-based process.

Morphological analysis performed by SEM showed a stabilizing effect of the polymer binder on the salt particles, while XRD data confirmed the presence of $SrBr_2 \cdot 6H_2O$ in the final material without undesired byproducts. The materials were characterized with different thermal analysis techniques to understand their performance in terms of energy storage density and capability of heat and mass transfer. The prepared composites were further molded in centimeter scale tabs suitable for exploitation in a modular-design reactor, in order to analyze their hydration kinetics and thermal conductivity properties. High contents of PDAC polyelectrolyte (SBH/G/P(1)) resulted in slightly limited dehydration kinetics and energy storage densities.

On the other hand, lower PDAC contents (SBH/G/P(0.1) or SBH/G/P(0.5)) did not affect dehydration kinetics and caused minimal reduction if energy storage density. Tabs prepared with SBH/G/P(0.5) were found to have significantly higher hydration rates in ambient conditions (23 °C and 50% RH) with respect to the conventional SBH/G composites. Indeed, the polyelectrolyte-containing formulation allowed us to reach complete hydration of the tabs in ~10 h, while the samples with no PDAC reached only ~35% of total hydration in the same time. These results relate to the physical action of the organic polyelectrolyte, acting as a binder between salt crystals, controlling moisture diffusion, and mechanically stabilizing the structure against stress-cracking, which is typical of pristine salt and salt/graphite formulations. Furthermore, a state of the art value of 16 W/mK thermal conductivity was obtained for the tabs, almost independent of the presence of the polyelectrolyte.

The results collected in this paper clearly demonstrate the proposed approach as a promising strategy for the design of efficient thermochemical storage solutions. Future studies should aim to investigate the cyclability of multiple hydration/dehydration cycles of the composite tabs, under controlled air flow, as well as their engineering in order to exploit processing conditions and geometries capable of reducing charge/discharge cycles and improving the efficiency of the system. Mechanical characterizations of the tabs might also prove important to better understand the stabilizing effect of the polyelectrolyte in the composite structure. All these characteristics will help in the fabrication of a TCM suitable for low grade heat reuse and ready for a scale up in prototypes focused on specific applications.

Supplementary Materials: The following are available online at http://www.mdpi.com/2079-4991/9/3/420/s1, Figure S1: Collected XRD diffractograms for (**a**) $SrBr_2 \cdot 6H_2O$ and, (**b**) expanded natural graphite. Figure S2: Thermal conductivity data of prepared composites.

Author Contributions: Conceptualization, G.S., A.F. and F.C.; methodology, S.S., F.C. and A.F.; validation, S.S.; formal analysis, S.S.; investigation, S.S.; data curation, S.S.; writing—original draft preparation, S.S.; writing—review and editing, F.C. and A.F.

Funding: This research received no external funding.

Acknowledgments: We would like to thank Francesco Cantamessa from Politecnico di Torino for help in the preliminary experiments in the development of composite materials. Daniele Battegazzore and Giusi Iacono from Politecnico di Torino are also gratefully acknowledged for XRD and SEM analyses. S.S. gratefully acknowledges the Center for Sustainable Future Technologies, Istituto Italiano di Tecnologia, for funding his PhD grant.

Conflicts of Interest: The authors declare no conflicts of interest. The funders had no role in the design of the study; in the collection, analyses, or interpretation of data; in the writing of the manuscript, or in the decision to publish the results.

References

1. Zhang, X. *Energy Solutions to Combat Global Warming*; Springer: New York, NY, USA, 2017; Volume 33, ISBN 978-3-319-26948-1.
2. Arce, P.; Medrano, M.; Gil, A.; Oró, E.; Cabeza, L.F. Overview of thermal energy storage (TES) potential energy savings and climate change mitigation in Spain and Europe. *Appl. Energy* **2011**, *88*, 2764–2774. [CrossRef]
3. Ammar, Y.; Joyce, S.; Norman, R.; Wang, Y.; Roskilly, A.P. Low grade thermal energy sources and uses from the process industry in the UK. *Appl. Energy* **2012**, *89*, 3–20. [CrossRef]
4. Bellocchi, S.; Leo, G.; Manno, M.; Pentimalli, M.; Salvatori, M.; Zaccagnini, A. Adsorbent materials for low-grade waste heat recovery: Application to industrial pasta drying processes. *Energy* **2017**, *140*, 729–745. [CrossRef]
5. Yesßiller, N.; Hanson, J.L.; Yee, E.H. Waste heat generation: A comprehensive review. *Waste Manag.* **2015**, *42*, 166–179. [CrossRef] [PubMed]
6. Wahlroos, M.; Pärssinen, M.; Rinne, S.; Syri, S.; Manner, J. Future views on waste heat utilization—Case of data centers in Northern Europe. *Renew. Sustain. Energy Rev.* **2018**, *82*, 1749–1764. [CrossRef]
7. Cot-Gores, J.; Castell, A.; Cabeza, L.F. Thermochemical energy storage and conversion: A-state-of-the-art review of the experimental research under practical conditions. *Renew. Sustain. Energy Rev.* **2012**, *16*, 5207–5224. [CrossRef]
8. Heier, J.; Bales, C.; Martin, V. Combining thermal energy storage with buildings—A review. *Renew. Sustain. Energy Rev.* **2015**, *42*, 1305–1325. [CrossRef]
9. Saffari, M.; de Gracia, A.; Fernández, C.; Belusko, M.; Boer, D.; Cabeza, L.F. Optimized demand side management (DSM) of peak electricity demand by coupling low temperature thermal energy storage (TES) and solar PV. *Appl. Energy* **2018**, *211*, 604–616. [CrossRef]
10. Islam, M.P.; Morimoto, T. Advances in low to medium temperature non-concentrating solar thermal technology. *Renew. Sustain. Energy Rev.* **2018**, *82*, 2066–2093. [CrossRef]
11. Sarbu, I. A Comprehensive Review of Thermal Energy Storage. *Sustainability* **2018**, *10*, 191. [CrossRef]
12. Scapino, L.; Zondag, H.A.; Van Bael, J.; Diriken, J.; Rindt, C.C.M. Sorption heat storage for long-term low-temperature applications: A review on the advancements at material and prototype scale. *Appl. Energy* **2017**, *190*, 920–948. [CrossRef]
13. Vasta, S.; Brancato, V.; La Rosa, D.; Palomba, V.; Restuccia, G.; Sapienza, A.; Frazzica, A. Adsorption Heat Storage: State of the Art and Future Perspectives. *Nanomaterials* **2018**, *8*, 522. [CrossRef] [PubMed]
14. Yu, N.; Wang, R.Z.; Wang, L.W. Sorption thermal storage for solar energy. *Prog. Energy Combust. Sci.* **2013**, *39*, 489–514. [CrossRef]
15. N'Tsoukpoe, K.E.; Schmidt, T.; Rammelberg, H.U.; Watts, B.A.; Ruck, W.K.L. A systematic multi-step screening of numerous salt hydrates for low temperature thermochemical energy storage. *Appl. Energy* **2014**, *124*, 1–16. [CrossRef]
16. Donkers, P.A.J. A review of salt hydrates for seasonal heat storage in domestic applications. *Appl. Energy* **2017**, *199*, 1–26. [CrossRef]
17. Richter, M.; Habermann, E.M.; Siebecke, E.; Linder, M. A systematic screening of salt hydrates as materials for a thermochemical heat transformer. *Thermochim. Acta* **2018**, *659*, 136–150. [CrossRef]
18. Kiyabu, S.; Lowe, J.S.; Ahmed, A.; Siegel, D.J. Computational Screening of Hydration Reactions for Thermal Energy Storage: New Materials and Design Rules. *Chem. Mater.* **2018**, *30*, 2006–2017. [CrossRef]
19. Fopah-Lele, A.; Tamba, J.G. A review on the use of $SrBr_2 \cdot 6H_2O$ as a potential material for low temperature energy storage systems and building applications. *Sol. Energy Mater. Sol. Cells* **2017**, *164*, 175–187. [CrossRef]

20. Fopah-Lele, A.; Rohde, C.; Neumann, K.; Tietjen, T.; Rönnebeck, T.; N'Tsoukpoe, K.E.; Osterland, T.; Opel, O.; Ruck, W.K. Lab-scale experiment of a closed thermochemical heat storage system including honeycomb heat exchanger. *Energy* **2016**, *114*, 225–238. [CrossRef]
21. Gordeeva, L.G.; Aristov, Y.I. Composites "salt inside porous matrix" for adsorption heat transformation: A current state-of-the-art and new trends. *Int. J. Low-Carbon Technol.* **2012**, *7*, 288–302. [CrossRef]
22. Korhammer, K.; Druske, M.-M.; Fopah-Lele, A.; Rammelberg, H.U.; Wegscheider, N.; Opel, O.; Osterland, T.; Ruck, W. Sorption and thermal characterization of composite materials based on chlorides for thermal energy storage. *Appl. Energy* **2016**, *162*, 1462–1472. [CrossRef]
23. Gaeini, M.; Rouws, A.L.; Salari, J.W.O.; Zondag, H.A.; Rindt, C.C.M. Characterization of microencapsulated and impregnated porous host materials based on calcium chloride for thermochemical energy storage. *Appl. Energy* **2018**, *212*, 1165–1177. [CrossRef]
24. Cammarata, A.; Verda, V.; Sciacovelli, A.; Ding, Y. Hybrid strontium bromide-natural graphite composites for low to medium temperature thermochemical energy storage: Formulation, fabrication and performance investigation. *Energy Convers. Manag.* **2018**, *166*, 233–240. [CrossRef]
25. Lu, J.; Do, I.; Fukushima, H.; Lee, I.; Drzal, L.T. Stable aqueous suspension and self-assembly of graphite nanoplatelets coated with various polyelectrolytes. *J. Nanomater.* **2010**, *2010*, 2. [CrossRef]
26. Qin, C.; Feng, Y.; An, H.; Han, J.; Cao, C.; Feng, W. Tetracarboxylated Azobenzene/Polymer Supramolecular Assemblies as High-Performance Multiresponsive Actuators. *ACS Appl. Mater. Interfaces* **2017**, *9*, 4066–4073. [CrossRef] [PubMed]
27. Donkers, P.A.J.; Pel, L.; Adan, O.C.G. Experimental studies for the cyclability of salt hydrates for thermochemical heat storage. *J. Energy Storage* **2016**, *5*, 25–32. [CrossRef]
28. Gustavsson, M.; Karawacki, E.; Gustafsson, S.E. Thermal conductivity, thermal diffusivity, and specific heat of thin samples from transient measurements with hot disk sensors. *Rev. Sci. Instrum.* **1994**, *65*, 3856–3859. [CrossRef]
29. JCPDS-ICDD. Available online: http://www.icdd.com/ (accessed on 10 December 2018).
30. Inoue, M.; Hirasawa, I. The relationship between crystal morphology and XRD peak intensity on $CaSO_4 \cdot 2H_2O$. *J. Cryst. Growth* **2013**, *380*, 169–175. [CrossRef]
31. Fopah, A.; Kuznik, F.; Opel, O.; Ruck, W.K.L. Performance analysis of a thermochemical based heat storage as an addition to cogeneration systems. *Energy Convers. Manag.* **2015**, *106*, 1327–1344. [CrossRef]
32. Zhao, Y.J.; Wang, R.Z.; Zhang, Y.N.; Yu, N. Development of $SrBr_2$ composite sorbents for a sorption thermal energy storage system to store low-temperature heat. *Energy* **2016**, *115*, 129–139. [CrossRef]

© 2019 by the authors. Licensee MDPI, Basel, Switzerland. This article is an open access article distributed under the terms and conditions of the Creative Commons Attribution (CC BY) license (http://creativecommons.org/licenses/by/4.0/).

Article

Cycle Stability and Hydration Behavior of Magnesium Oxide and Its Dependence on the Precursor-Related Particle Morphology

Georg Gravogl [1,2], Christian Knoll [2,3], Jan M. Welch [4], Werner Artner [5], Norbert Freiberger [6], Roland Nilica [6], Elisabeth Eitenberger [7], Gernot Friedbacher [7], Michael Harasek [3], Andreas Werner [8], Klaudia Hradil [5], Herwig Peterlik [9], Peter Weinberger [2], Danny Müller [2,*] and Ronald Miletich [1]

1. Department of Mineralogy and Crystallography, University of Vienna, Althanstraße 14, 1090 Vienna, Austria; georg.gravogl@tuwien.ac.at (G.G.); ronald.miletich-pawliczek@univie.ac.at (R.M.)
2. Institute of Applied Synthetic Chemistry, TU Wien, Getreidemarkt 9, 1060 Vienna, Austria; christian.knoll@tuwien.ac.at (C.K.); peter.e163.weinberger@tuwien.ac.at (P.W.)
3. Institute of Chemical, Environmental & Biological Engineering, TU Wien, Getreidemarkt 9, 1060 Vienna, Austria; Michael.harasek@tuwien.ac.at
4. Atominstitut, TU Wien, Stadionallee 2, 1020 Vienna, Austria; jan.welch@tuwien.ac.at
5. X-ray Center, TU Wien, Getreidemarkt 9, 1060 Vienna, Austria; werner.artner@tuwien.ac.at (W.A.); Klaudia.hradil@tuwien.ac.at (K.H.)
6. RHI-AG, Magnesitstraße 2, 8700 Leoben, Austria; Norbert.freiberger@rhimagnesita.com (N.F.); Roland.nilica@rhimagnesita.com (R.N.)
7. Institute of Chemical Technologies and Analytics, TU Wien, Getreidemarkt 9, 1060 Vienna, Austria; Elisabeth.eitenberger@tuwien.ac.at (E.E.); Gernot.friedbacher@tuwien.ac.at (G.F.)
8. Institute for Energy Systems and Thermodynamics, TU Wien, Getreidemarkt 9, 1060 Vienna, Austria; Andreas.werner@tuwien.ac.at
9. Faculty of Physics, University of Vienna, Boltzmanngasse 5, 1090 Vienna, Austria; Herwig.peterlik@univie.ac.at
* Correspondence: danny.mueller@tuwien.ac.at; Tel.: +43-1-5880-1163-740

Received: 31 August 2018; Accepted: 2 October 2018; Published: 7 October 2018

Abstract: Thermochemical energy storage is considered as an auspicious method for the recycling of medium-temperature waste heat. The reaction couple $Mg(OH)_2$–MgO is intensely investigated for this purpose, suffering so far from limited cycle stability. To overcome this issue, $Mg(OH)_2$, $MgCO_3$, and $MgC_2O_4 \cdot 2H_2O$ were compared as precursor materials for MgO production. Depending on the precursor, the particle morphology of the resulting MgO changes, resulting in different hydration behavior and cycle stability. Agglomeration of the material during cyclization was identified as main reason for the decreased reactivity. Immersion of the spent material in liquid H_2O decomposes the agglomerates restoring the initial reactivity of the material, thus serving as a regeneration step.

Keywords: particle morphology; magnesium hydroxide; magnesium carbonate; magnesium oxalate; magnesium oxide; cycle stability; in-situ powder X-ray diffraction (PXRD); hydration reactivity; thermochemical energy storage; thermochemistry

1. Introduction

Energy management is a major challenge for our society, requiring equal measures of political and scientific involvement [1]. Energy supply, sustainable, environmentally benign energy production, and efficient utilization are key issues in managing global energy use [2]. Energy management may, in many cases, be better expressed as 'heat management', as heat is the most ubiquitous form of energy.

In nearly all types of electrical power plants, as well as in most industrial processes, heat is used as the driving force and operating medium. Within this context, the utilization of waste heat, accounting for two-thirds of overall global energy production, is an extensively investigated field [3]. The use of waste heat flows includes several aspects, one of them being temporal decoupling of waste heat availability and demand, as the two are not necessarily correlated. The necessary storage may be realized using materials for sensible, latent, or thermochemical storage of energy (heat) [4–9]. All three energy storage concepts offer advantages in specific areas of application [6,9,10].

Thermochemical energy storage (TCES) features long-term storage, a wide range of compatible temperatures, applicability as a heat pump system, and finally, high energy storage densities [10–13]. Based on these aspects, medium-temperature waste heat (up to 450 °C and extensively available from industrial processes) is perfectly suitable for TCES systems. An attractive TCES material for medium-temperature applications is the system $Mg(OH)_2$–MgO with a storage temperature around 350 °C [14]. Both $Mg(OH)_2$ and MgO are industrial base materials and are, therefore, available in large quantities at low prices.

$Mg(OH)_2$–MgO as a TCES material is well known for this purpose, with many aspects related to its application in energy storage already investigated in literature. Kinetic investigations of dehydration and rehydration [15,16], mechanistic aspects of the conversion [17,18], modification of the material by additions of lithium salts [16,19,20], by coating or use of composite material [21,22], by dotation [23], and finally also applicability in form of a chemical heat pump [24] were reported. Nonetheless, two key issues preventing industrial application remain unaddressed: First, rehydration reactivity (completeness), and second, the cycle stability. Whereas for the limited cyclability observed thus far, no satisfying solution has been found, the rehydration reactivity is addressed by the addition of lithium salts [16,19,20], which are quite expensive. On a molecular level, reactivity could also be tuned by dotation of $Mg(OH)_2$ with Ca^{2+}-ions [23].

On an industrial scale MgO is produced via calcination of $Mg(OH)_2$ or $MgCO_3$ [25]. Both precursors are found in natural deposits, but whereas $MgCO_3$ is an industrially mined raw material, $Mg(OH)_2$ is produced from serpentinite or processing seawater [26]. However, aerobic calcination of any other Mg compound may result in formation of MgO by stepwise decomposition. In Scheme 1, this is shown at the example of the mentioned industrial precursor, as well as for magnesium oxalate dihydrate.

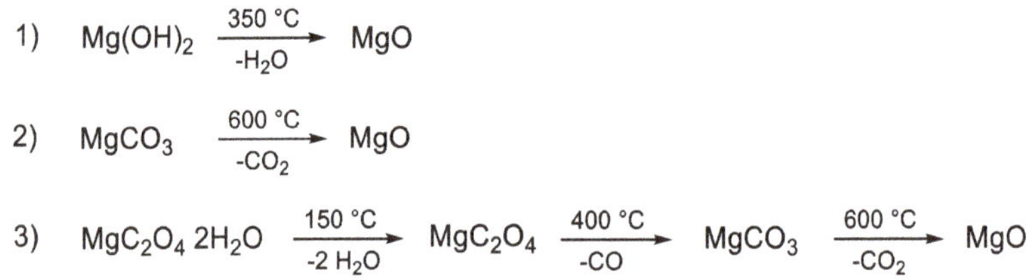

1) $Mg(OH)_2 \xrightarrow[-H_2O]{350\ °C} MgO$

2) $MgCO_3 \xrightarrow[-CO_2]{600\ °C} MgO$

3) $MgC_2O_4 \cdot 2H_2O \xrightarrow[-2\ H_2O]{150\ °C} MgC_2O_4 \xrightarrow[-CO]{400\ °C} MgCO_3 \xrightarrow[-CO_2]{600\ °C} MgO$

Scheme 1. Thermal decomposition of various MgO precursors: (1) $Mg(OH)_2$, (2) $MgCO_3$, (3) $MgC_2O_4 \cdot 2H_2O$.

All so far performed investigations on the rehydration of MgO for thermochemical energy storage purposes have largely neglected the origin of the MgO. As $Mg(OH)_2$, $MgCO_3$, $MgC_2O_4 \cdot 2H_2O$, and MgO crystallize in crystallographically and stereochemically different systems (Table 1), and feature notably different particle morphologies, MgO samples originating from different precursors can not necessarily be expected to have the same properties with respect to rehydration and cycle stability. This assumption is supported by previous kinetic studies on the H_2O-dissociation on MgO. Compared to the isotypic CaO [27], the lower hydration reactivity of MgO [28] is mainly caused by the kinetic barrier of the water dissociation on the surface [29]. The disfavored H_2O-dissociation as first step in formation of $Mg(OH)_2$ occurs mainly at surface defects, edges, step edges, or corner sites, exhibiting a lower dissociation energy barrier [30]. This suggests that by variation of the particle morphology and origin of the MgO, the rehydration behavior should be affected. While all precursors result in compositionally indistinguishable MgO sample stoichiometries, the particle size and morphology, crystallographic orientation, and thus the orientation of the reactive surfaces of the material are not necessarily the same. To verify this hypothesis, MgO obtained by calcination of $Mg(OH)_2$, $MgCO_3$, and $MgC_2O_4 \cdot 2H_2O$ was investigated regarding hydration reactivity and cycle stability.

Table 1. Comparison of the crystallographic parameters of selected MgO precursors and MgO.

	$Mg(OH)_2$ [31]	$MgCO_3$ [32]	$MgC_2O_4 \cdot 2H_2O$ [33]	MgO [34]
Space group	$P\bar{3}m1$ (164)	$R\bar{3}c$ (167)	Fddd (70)	$Fm\bar{3}m$ (225)
a [Å]	3.1486(1)	4.637(1)	5.3940(11)	4.2113(5)
b [Å]	3.1486(1	4.637(1)	12.691(3)	4.2113(5)
c [Å]	4.7713(1)	15.023(3)	15.399(3)	4.2113(5)
α [°]	90	90	90	90
γ [°]	120	120	90	90
V [Å^3]	40.96	279.74	1054.14	74.69

2. Materials and Methods

2.1. Material

$Mg(OH)_2$ powder (particle size ≤5 μm) and $MgCO_3$ (particle size ≤200 μm) were supplied by RHI-AG (X-ray fluorescence analysis (Bruker AXS GmbH, 76187 Karlsruhe, Germany)) of the materials revealed no significant impurities). $MgC_2O_4 \cdot 2H_2O$ (98.5% purity) was purchased from abcr (GmBH, 76187 Karlsruhe, Germany) and the particle fraction ≤200 μm was used as supplied. The materials were calcined in an electric furnace under air and a static atmosphere for 4 h at variable temperatures ($Mg(OH)_2$: at 375 °C; $MgCO_3$: at 550 °C, 600 °C, 650 °C; $MgC_2O_4 \cdot 2H_2O$: at 650 °C). For subsequent rehydration, the in-situ calcined material from the (powder X-ray diffraction) P-XRD measurement was kept for 24 h in liquid water under ambient pressure-temperature conditions.

2.2. BET Surface

The specific surface of the samples was determined by nitrogen sorption measurements, which were performed on an ASAP 2020 (Micromeritics) instrument. The samples (amounting between 100–200 mg) were degassed under vacuum at 80 °C overnight prior to measurement. The surface area was calculated according to Brunauer, Emmett, and Teller (BET, Micromeritics Instrument Corp., Norcross, GA, USA) and t-plot methods [35].

2.3. Powder X-ray Diffraction with In-Situ Hydration (P-XRD)

Hydration of calcined samples was performed in an Anton Paar XRK 900 (Bruker AXS GmbH, 76187 Karlsruhe, Germany) sample chamber, connected to an evaporation coil kept at 300 °C (see Figure S1a). Using an HPLC-pump, water was evaporated at rates from 1 g H_2O min^{-1} up to 3 g min^{-1} and the resulting steam was passed through the sample (1 mm thickness) with 0.2 L min^{-1} nitrogen as carrier gas. The sample is mounted on a hollow ceramic powder sample holder, allowing for complete perfusion of the sample with the water vapour (see Figure S1b). As the sample is completely penetrated by the X-rays, the obtained diffractograms represent an average across the total sample with respect to the quantitative phase proportions. The diffractograms were evaluated using the PANalytical program suite HighScorePlus v3.0d. A background correction and a $K_{\alpha 2}$ strip were performed. Phase assignment is based on the ICDD-PDF4+ database (International Diffraction Data-Powder Diffraction File), the exact phase composition, shown in the conversion plots, was obtained via Rietveld-refinement [36] in the program suite HighScorePlus v3.0d. All quantifications based on P-XRD are accurate within of ±2%. The rehydration rates were calculated based on the phase composition derived from the diffractograms, normalizing the percentages of Mg(OH)$_2$ and MgO to a total of 100%.

2.4. Scanning Electron Microscopy (SEM)

SEM (Thermo Fisher Scientific, 168 Third Avenue, Waltham, MA 02451, USA)) images were recorded on gold coated samples with a Quanta 200 SEM instrument from FEI under low-vacuum at a water vapour pressure of 80 Pa to prevent electrostatic charging.

2.5. Small-Angle X-ray Scattering (SAXS)

The samples were prepared either as powder between two pieces of tape or in a sealed capillary. Patterns were recorded using a microsource with X-rays from a copper target (Incoatec High Brilliance, wavelength 0.1542 nm, CuK$_\alpha$), a point focus (Nanostar from Bruker AXS) and a 2D detector (VÅNTEC 2000). The X-ray patterns were radially averaged and background corrected to obtain scattering intensities in dependence on the scattering vector $q = (4\pi/\lambda)\sin\theta$, with 2θ being the scattering angle.

The fit function from *Beaucage* [37] to describe scattering intensities of complex systems with a broad size distribution consists of a power law and Guinier's exponential form,

$$I(q) \propto G \exp\left(\frac{-q^2 R_g^2}{3}\right) + B \left[\frac{\left(erf(qR_g/\sqrt{6})\right)^3}{q}\right]^{d_f} \quad (1)$$

where G and B are the numerical prefactors, d_f is the fractal dimension, R_g is the radius of gyration and $erf(x)$ is the error function. To describe the particle interference and thus the tendency of particles to agglomerate, additionally a structure factor from a hard sphere model was used [38,39],

$$I(q) \propto \left(G \exp\left(\frac{-q^2 R_g^2}{3}\right) + B \left[\frac{\left(erf(qR_g/\sqrt{6})\right)^3}{q}\right]^{d_f}\right) S(q) \quad (2)$$

with

$$S(q) = 1/(1 + 24\eta \, G_{int}(2qR_{HS})/(2qR_{HS})) \quad (3)$$

and R_{HS} being the hard sphere radius describing a typical distance of objects, η the hard sphere volume factor for characterizing the amount of agglomeration, and G_{int} a function derived in Kinning et al. [38].

3. Discussion and Results

To combine the apparent particle morphology with the crystallographic features of the lattice as given in Table 1, SEM-images of the original and calcined materials are compared in Figure 1. The first row corresponds to SEM-images of the various MgO precursors; in the second row the resulting MgO samples, obtained after thermal decomposition, are shown. Whereas Mg(OH)$_2$ particles feature euhedral idiomorphic shapes with characteristic faces following hexagonal symmetry (Figure 1a), both MgCO$_3$ (Figure 1b) and MgC$_2$O$_4$·2H$_2$O (Figure 1c) reveal subidiomorphic irregular particle shapes occasionally showing typical rhombohedral (Figure 1b) or foliated (Figure 1c) cleavage faces, which in the case of MgC$_2$O$_4$·2H$_2$O correspond to its layer structure.

The particle morphology of the materials changes during calcination (Figure 1, second row), leading to three differently textured MgO samples. Whereas for using Mg(OH)$_2$ as the precursor material (Figure 1a), calcination results in an apparently unchanged particle morphologies, the MgO crystallites obtained from both MgCO$_3$ (Figure 1b) and MgC$_2$O$_4$·2H$_2$O (Figure 1c) precursors are characterized by a clear surface fragmentation, which can be attributed to larger degree of structural reconstruction on the release of volatile components. In contrast, the H$_2$O release from Mg(OH)$_2$ to

MgO follows a simple change from hcp to ccp arrangement of the octahedral subunits and hence preserves the particles in its shape to a large extent.

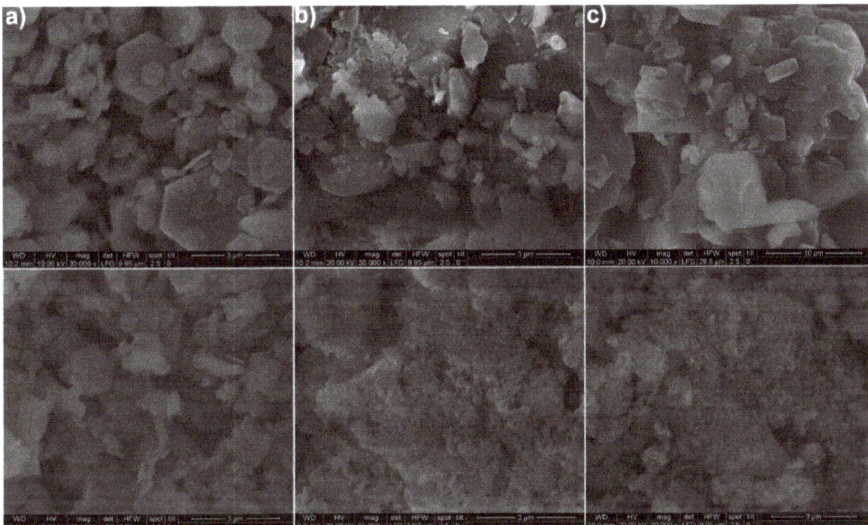

Figure 1. SEM pictures of (**a**) Mg(OH)$_2$; (**b**) MgCO$_3$; (**c**) MgC$_2$O$_4$·2H$_2$O before calcination (first row) and after calcination (second row).

On the nanoscale, small-angle X-ray scattering (SAXS) reveals a transformation of the material from a dense solid to a highly porous material on calcination (see Figure S2). The nanostructure of MgO was modelled by a unified Guinier/power law [37], resulting in a radius of gyration for the size of the particles and an agglomeration with a structure factor from a hard sphere model, describing the agglomeration of particles with a typical distance $2R_{HS}$ and the packing density with a hard sphere volume ratio η [38,39]. The detailed fit parameters are found in the supporting information (Table S1). In general, the gyration radius of MgO particles calcined from MgCO$_3$ and MgC$_2$O$_4$·2H$_2$O is about 6.6 and 5.1 nm, respectively, in comparison to about 2 nm if calcined from Mg(OH)$_2$. In contrast, the values of η = 0.18 and a fractal dimension of d_f = 2.8 indicate, that MgO from Mg(OH)$_2$ consists of small, agglomerated particles with a wide size distribution, whereas MgO from other precursors is built up of larger, denser nanoparticles (η close to zero, d_f = 4).

In order to allow a better comparability between the different precursor materials investigated within this study, MgO obtained by calcination from Mg(OH)$_2$ was used as reference material [28]. In Scheme 2, schematic representation of the calcination and rehydration conditions applied for its preparation is shown.

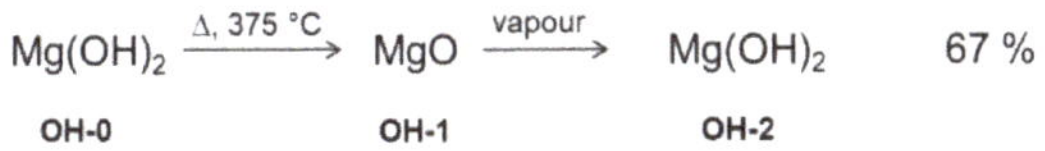

$$Mg(OH)_2 \xrightarrow{\Delta,\ 375\ °C} MgO \xrightarrow{vapour} Mg(OH)_2 \quad 67\ \%$$
$$\text{OH-0} \qquad\qquad \text{OH-1} \qquad\qquad \text{OH-2}$$

Scheme 2. Conditions for calcination and rehydration of MgO obtained from Mg(OH)$_2$.

To correlate the particle morphologies with rehydration reactivity and cycle stability, rehydration experiments using the different MgO samples were monitored by in situ powder X-ray diffraction (P-XRD). This allows for a direct observation and quantification of the reaction progress. As in previous experiments, the rehydration reactivity of the MgO produced from Mg(OH)$_2$ was found quite limited [28]. To eventually increase the reaction rate, an even larger excess of water vapour was introduced into the reaction chamber. Increasing the vapour flow from 1 g min^{-1} to 3 g min^{-1}

enhanced the rehydration conversion of MgO from 44% to 67% (Figure 2a). To assess the cycle stability for the increased vapour flow, five consecutive rehydration–calcination cycles were performed (Figure 2b). Similar to previous experiments with lower vapour flows the rehydration yield decreased over 5 cycles to a final Mg(OH)$_2$ conversion of only 14%. Even after the first cycle the rehydration conversion was depleted to 43%.

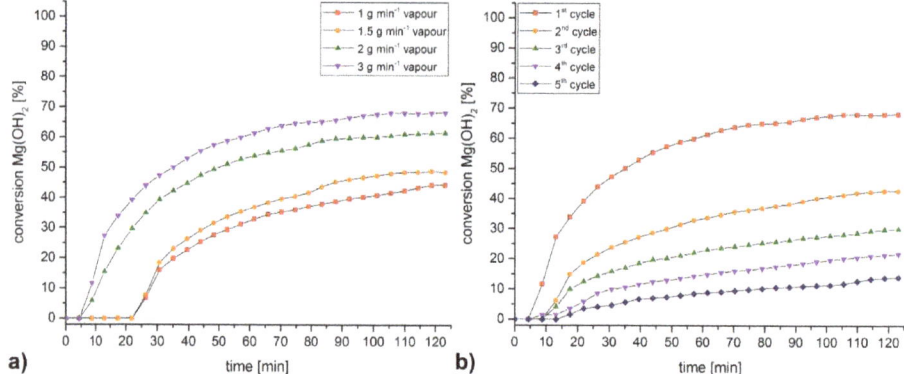

Figure 2. (a) Rehydration of Mg(OH)$_2$-originating MgO with various water vapour flow rates; (b) Cycle stability of Mg(OH)$_2$-originating MgO.

For MgO produced by calcination of Mg(OH)$_2$ a strong correlation between reactivity, accessible surface area and calcination temperature has been established [28]. Higher calcination temperatures promote sintering of the particles, leading to a decreased porosity, increased MgO crystal size, and decreased rehydration yield.

To assess the possibility of a similar effect for MgO originating from MgCO$_3$, initial studies of the correlation between calcination temperature, BET surface and rehydration reactivity were made. For this purpose, samples of MgCO$_3$ were calcined at 550 °C, 600 °C, and 650 °C for 3, 6, 9, and 12 h. The MgO formed from MgCO$_3$ calcined for 6 h at 600 °C had the highest surface area (Figure S3). Nevertheless, attempted rehydration of all samples by water vapour in the P-XRD failed, showing no Mg(OH)$_2$ formation within 120 minutes. In Scheme 3, representation of the conditions applied for calcination and rehydration of MgCO$_3$-derived MgO is given.

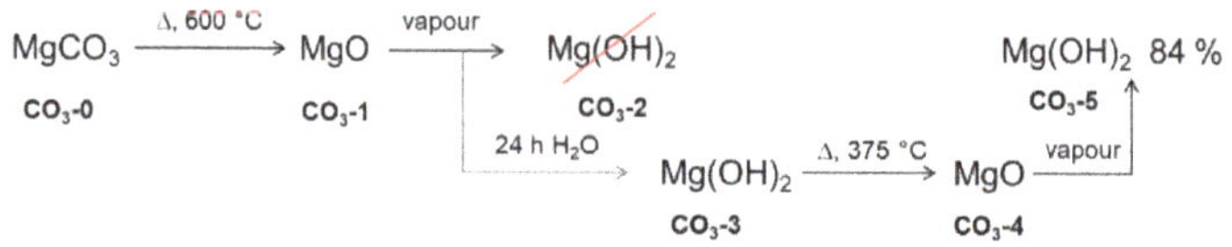

Scheme 3. Conditions for calcination and rehydration of MgO obtained from MgCO$_3$.

Based on the assumption, that a different chemical history of the MgO would have an impact on the reactivity during rehydration, a varied rehydration rate would have been expected. Observing no conversion to Mg(OH)$_2$ under the applied conditions was, however, quite unexpected. On prolonged exposure to water vapour over 24 h for MgO **CO$_3$-1** (see Scheme 3) a very sluggish formation of Mg(OH)$_2$ below 10% was observed. To ascertain whether rehydration of this material could be driven by longer exposure to a vast excess of reactant, the samples were stored in liquid water. After 24 h reaction time in water at room-temperature, according to P-XRD measurements the material had been completely transformed to Mg(OH)$_2$ (**CO$_3$-3**) To repeat the in situ rehydration study for this material, the calcination step was repeated at 375 °C using the conditions developed for Mg(OH)$_2$ [28]. After a new calcination step the BET surface of the various samples **CO$_3$-4** was found to be slightly higher than for **CO₃-1**, the MgO originating directly from MgCO$_3$ (Figure S4). In contrast to the first attempt, now for those materials the rehydration experiments in the P-XRD were repeated successfully for all materials (for detailed rehydration rates see Figure S5). Ranked according to their final conversion to Mg(OH)$_2$, the most reactive material within this series was obtained by calcination of MgCO$_3$ at 600 °C for 6 h (Figure 3) and subsequent rehydration in liquid water. A final conversion to 84% Mg(OH)$_2$ was not only by far the highest yield for the MgCO$_3$-originating series, but also notably more than for MgO originating from Mg(OH)$_2$ (67% final conversion).

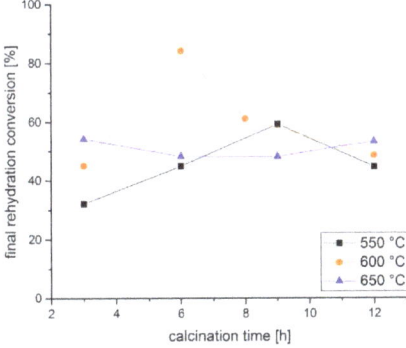

Figure 3. Final conversion for rehydration of the various MgCO$_3$-originating MgO samples **CO₃-4** in the P-XRD.

SEM images demonstrate, that the particle morphology of MgCO$_3$ (Figure 4a) is retained after calcination, although the formerly distinct edges and surfaces are now covered by smaller scales (Figure 4b). During the hydration of the calcined material in liquid water the large particles disintegrate into smaller platelets (Figure 4c), although lacking the characteristic hexagonal morphology as characteristic for euhedrally grown Mg(OH)$_2$ (see Figure 1). A subsequent calcination of material rehydrated in liquid water retains the afore mentioned platelet morphology (Figure 4d).

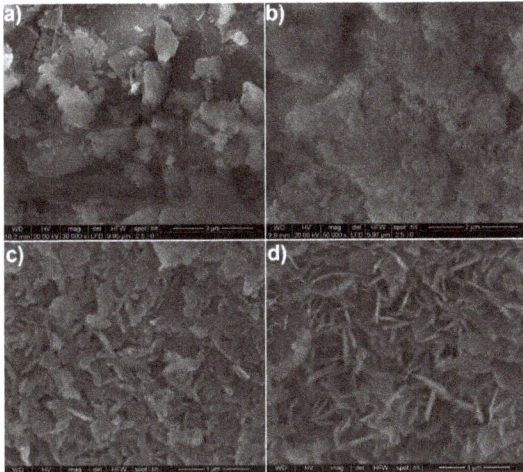

Figure 4. SEM images of (**a**) MgCO$_3$, (**b**) calcined MgCO$_3$ (**CO₃-1**), (**c**) calcined MgCO$_3$, rehydrated for 24 h in liquid water (**CO₃-3**), (**d**) material from image c after calcination (**CO₃-4**).

The changing particle shape observed in the SEM images is attributed to the volume work going along with the Mg(OH)$_2$ formation. To enable this rehydration-related rearrangement of the material, water in its liquid form seems crucial as an agent triggering the rehydration process. SAXS intensities (Figure S6) show that on (repeated) rehydration of the MgCO$_3$-derived MgO-samples the particle morphology is widely unchanged. From the larger scattering intensity a highly porous nanostructure, retained during rehydration, may be extrapolated. At the same time, a general decrease in particle size was also observed, being in good agreement with the SEM images (Figure 4).

The carbonate-derived MgO **CO$_3$-4** was also investigated in terms of cycle stability (Figure 5). Similar to the material **OH-1** originating from Mg(OH)$_2$, also in the case of **CO$_3$-4** a decrease in rehydration reactivity was detected, although to a lesser extent than observed for **OH-1**. Over five cycles the rehydration conversion drops to 57% (84% in the 1st, 75% in the 2nd cycle).

Figure 5. Cycle stability of MgCO$_3$-originating MgO **CO$_3$-4**.

As a third precursor for preparation of reactive MgO, MgC$_2$O$_4$·2H$_2$O was investigated (see Scheme 4).

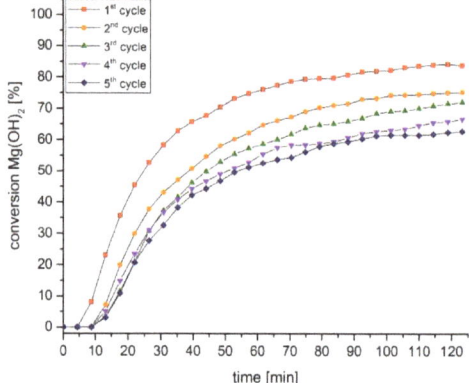

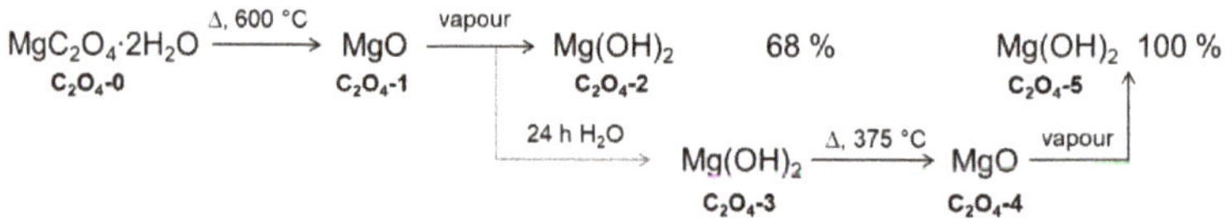

Scheme 4. Conditions for calcination and rehydration of MgO obtained from MgC$_2$O$_4$·2H$_2$O.

Since MgC$_2$O$_4$·2H$_2$O decomposes stepwise via MgCO$_3$ (see Scheme 1), only samples calcined in the furnace at 600 °C for 6 h were investigated. A comparison of the SEM images in Figure 6, compares the morphology of the different samples: initial oxalate material **C$_2$O$_4$-0** (Figure 6a), the calcined material **C$_2$O$_4$-1** (Figure 6b), MgO after rehydration in liquid water **C$_2$O$_4$-3** (Figure 6c), and a new calcined material **C$_2$O$_4$-4** (Figure 6d). Similar to the MgCO$_3$-case, calcination of the initial material resulted in partial fragmentation, whereas subsequent treatment with liquid water and re-calcination forced the material to adopt a lamellar-structured particle morphology. In contrast to the MgO originating from MgCO$_3$, thinner platelets were formed, those structure is preserved after calcination. Moreover, even the rehydration of the oxalate-based MgO did not yield the typical hexagonally shaped morphologies of euhedral brucite crystallites.

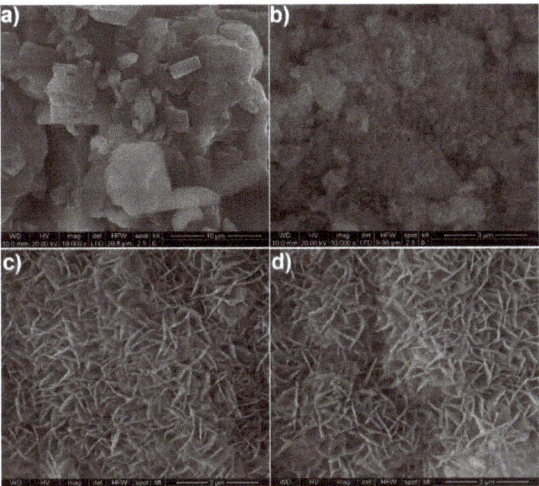

Figure 6. SEM images of (**a**) $MgC_2O_4 \cdot 2H_2O$, (**b**) calcined $MgC_2O_4 \cdot 2H_2O$ (**C_2O_4-1**), (**c**) calcined $MgC_2O_4 \cdot 2H_2O$, rehydrated for 24 h in liquid water (**C_2O_4-3**), (**d**) material from image c after calcination (**C_2O_4-4**).

Both SEM and SAXS data show a comparable picture as observed for $MgCO_3$. Rehydration of MgO **C_2O_4-1** in liquid water results in conservation of the original particle morphology to a wide extent (Figure S7). The porous nanostructure formed is preserved during rehydration. Accordingly, the increased SAXS intensities observed for the rehydrated material seem to arise from the formation of smaller particles (SEM images, Figure 6).

To determine the reactivity of the $MgC_2O_4 \cdot 2H_2O$-based MgO, both the material obtained by direct calcination of $MgC_2O_4 \cdot 2H_2O$ (**C_2O_4-1**) and that resulting from the rehydration–calcination sequence (**C_2O_4-4**) were subject to rehydration in the P-XRD. Unlike the case of $MgCO_3$, the directly calcined material **C_2O_4-1** was found to be reactive to rehydration, resulting in a final conversion of 68% (Figure 7a).

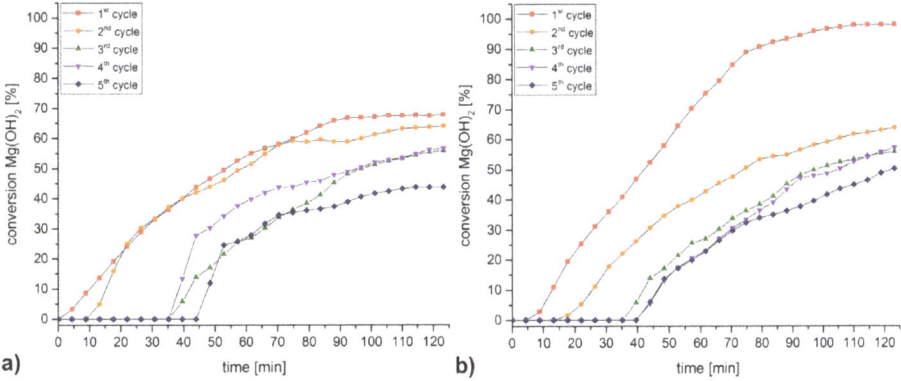

Figure 7. (**a**) Direct rehydration and cycle stability of MgC_2O_4-originating MgO **C_2O_4-1**; (**b**) Rehydration and cycle stability of MgC_2O_4-originating MgO **C_2O_4-4**.

The most remarkable finding is, that several batches of **C_2O_4-4**, the sample previously rehydrated in liquid H_2O and subsequently calcined, could be fully rehydrated in the 1st cycle, but in the 2nd cycle the rehydration conversion decreased to 64% (Figure 7b). After five cycles both samples gave a comparable final conversion of slightly less than 50% Mg(OH)$_2$.

Assessing the technological feasibility of MgC$_2$O$_4 \cdot$2H$_2$O as precursor material, on the one hand the material was completely rehydrated to Mg(OH)$_2$ within the first cycle after an initial treatment with liquid water. On the other hand, due to a large decrease in conversion rate during the successive cycles, a modest overall performance, and a relatively higher price compared to Mg(OH)$_2$ and MgCO$_3$, MgC$_2$O$_4 \cdot$2H$_2$O is most likely not suitable as a competitive MgO precursor. Therefore, MgC$_2$O$_4 \cdot$2H$_2$O was not subjected further studies.

Based on the conversion-enhancing effect of rehydrating calcined material in liquid water, both a sample of spent MgO originating from Mg(OH)$_2$ and one from MgCO$_3$ after the 5th rehydration cycle was calcined a further time and then rehydrated for 24 h in liquid water. At that point, P-XRD analysis showed complete transformation to Mg(OH)$_2$ for both samples. Both Mg(OH)$_2$-samples were now subjected further five rehydration–calcination cycles in the P-XRD, followed by a further regeneration in liquid water and another five rehydration–calcination cycles in the P-XRD. The conversion rates for 15 consecutive cycles, including two regeneration steps after five cycles are shown in Figure 8.

In the case of MgO originating from Mg(OH)$_2$, shown in Figure 8a, the conversion rate after the first regeneration (Cycle 6) was slightly enhanced compared to the very 1st cycle. This effect was even more pronounced after the second regeneration (Cycle 10), revealing an even further increased reactivity. Nevertheless, the depletion evidenced in the second cycle was retained even after the second cycles after regeneration, as observed for Cycles 6 and 12. These results could be reproduced on various batches.

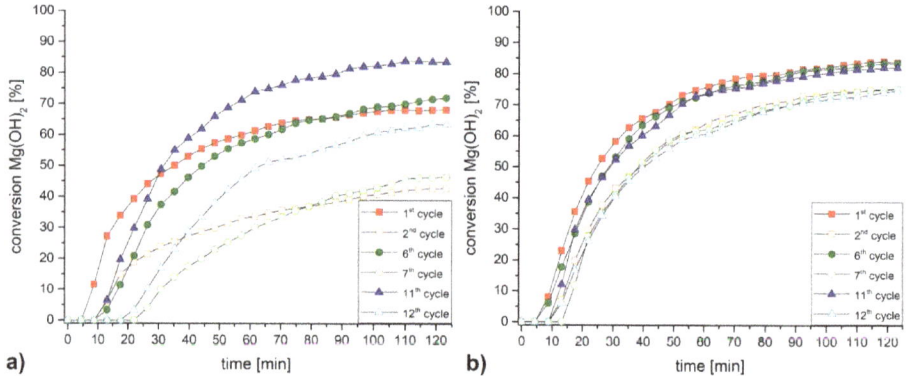

Figure 8. Selected conversion rates from a series of 15 consecutive calcination–hydration cycles, including two regeneration steps in liquid water after the 5th and the 10th cycle. (**a**) Mg(OH)$_2$-originating MgO after regeneration; (**b**) MgCO$_3$-originating MgO after regeneration.

In the case of MgO derived from MgCO$_3$, a different but even more promising effect was observed: The spent material could be completely regenerated to reproduce the reactivity observed for the first cycle. Even the second cycle after each regeneration process (Cycle 7 and 12) was comparable to the "first" second cycle. Even in this case the effect was reproducible on various batches.

To better understand the physical processes during regeneration, SEM images of material during several stages of regeneration and cycling were compared. Despite differences in initial particle morphology (Figure 9a,d), MgO originating from Mg(OH)$_2$ or MgCO$_3$ shows similar evolution of reactivity during repeated calcination–rehydration cycles. After five consecutive cycles, resulting in aged material of depleted rehydration reactivity (see Figures 2 and 5), the particle morphology of Mg(OH)$_2$-derived MgO (Figure 9b) seems nearly unaffected. In contrast, for material originating from MgCO$_3$, the larger spherical aggregates are retained (Figure 9e). After regeneration for 24 h in liquid water both materials reveal a lamellar, platelet morphology devoid of the characteristic hexagonal brucite particle shape (Figure 9c,f).

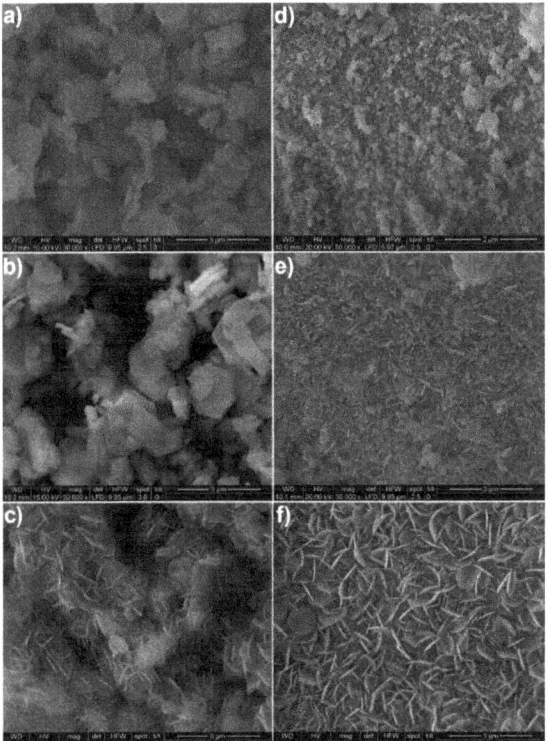

Figure 9. SEM images of various intermediates during calcination/rehydration/regeneration for Mg(OH)$_2$- (left) and MgCO$_3$-originating MgO (right); (**a**) Mg(OH)$_2$-originating MgO; (**b**) Mg(OH)$_2$-originating MgO after 5 rehydration–calcination cycles; (**c**) material of image b after regeneration for 24 h in liquid H$_2$O; (**d**) MgCO$_3$-originating MgO; (**e**) MgCO$_3$-originating MgO after five rehydration–calcination cycles; (**f**) material of image **e** after regeneration for 24 h in liquid H$_2$O.

To directly monitor the rehydration/regeneration process, both a sample of MgO originating from Mg(OH)$_2$ and from MgCO$_3$ were observed during rehydration in liquid water by in situ SAXS (Figures S8 and S9). Initial SAXS curves (black) and final SAXS curves (red) are highlighted to better visualize the data and clearly show the change in the inner structure of the particles. The main difference is the time required for structural recovery, which is about three-fold shorter for Mg(OH)$_2$-derived material compared to that from MgCO$_3$ (Figure S8), most likely due to the considerably larger particle sizes favored by the latter.

Within various repeated experiments the regeneration process described for spent MgO was found to be reproducible—on one hand for the material of the same origin, on the other for materials of different origins. We suggest that during the regeneration process due to the comparably long reaction time and the vast excess of water, a complete conversion to Mg(OH)$_2$ as well as a regeneration of the particle morphology occurs. Both effects complement each other, restoring the original reactivity of the material.

The possibility of a regeneration of spent material is of utmost importance for assessing the economic feasibility of a TCES material and energy storage process, as by prolonging the life-time the materials investment costs are minimized. Additionally, by implementation of a continuous

4. Conclusions

MgO obtained by calcination of $Mg(OH)_2$, $MgCO_3$, and $MgC_2O_4 \cdot 2H_2O$ was compared regarding its rehydration reactivity and cycle stability to assess its applicability in thermochemical energy storage. The three different MgO-precursors led to three MgO samples featuring different particle morphologies with identical chemical compositions. Whereas $Mg(OH)_2$ and $MgC_2O_4 \cdot 2H_2O$ resulted in reactive MgO that could be rehydrated by water vapour to $Mg(OH)_2$ directly following calcination, material originating from $MgCO_3$ resulted in no conversion on contact with water vapour. Only after rehydration in liquid water and subsequent calcination of the thus formed $Mg(OH)_2$, 84% of the resulting material could be rehydrated by water vapour. All materials investigated showed decreased rehydration reactivity during consecutive calcination–rehydration cycles, with $MgCO_3$-derived MgO showing the smallest decline in reactivity. A regeneration step, consisting of rehydration of the spent material in liquid water over 24 h, restored the initial reactivity allowing for recycling of the material. In the case of $Mg(OH)_2$ derived material, the initial reactivity could even be improved by repeated regeneration of the material in liquid water.

The results reported herein confirm, that the reactivity of MgO towards rehydration is strongly correlated to origin and physicochemical history of the material–an aspect so far neglected in the research on TCES materials. The correlation between chemical history and performance of storage materials may stimulate additional to coating, chemical dotation, etc., the consideration of a further, easily tunable parameter for the research on novel TCES systems.

Supplementary Materials: The following are available online at http://www.mdpi.com/2079-4991/8/10/795/s1, Figure S1: Rehydration setup, reaction chamber and sample holder used for the in situ studies. Figure S2: SAXS intensities of starting materials and materials after calcination. Figure S3: BET surfaces of the $MgCO_3$-originating MgO samples. Figure S4: BET surfaces of the $MgCO_3$-originating MgO samples after rehydration. Figure S5: Rehydration rates of $MgCO_3$-originating MgO samples in the P-XRD. Figure S6: SAXS intensities of materials from $MgCO_3$ precursor. Figure S7: SAXS intensities of materials from $Mg_2C_2O_4 \cdot 2H_2O$ precursor. Figure S8: In situ SAXS intensities during regeneration in liquid water for 24 h. Figure S9: Kinetics of conversion to hydroxide during regeneration in liquid water. Table S1: Fit data for calcined materials

Author Contributions: Experimental investigation: G.G., C.K.; Proof-reading and language: J.M.W.; Evaluation of P-XRD data: W.A., K.H.; Provision of samples and scientific contribution: N.F., R.N.; SEM images and interpretation: E.E., G.F.; SAXS measurements and interpretation: H.P.; Project administration: A.W.; Conception of the study, writing, review, and editing: D.M.; Supervision and funding acquisition: M.H., P.W., R.M.

Funding: This research was funded by the Austrian Research Promotion Agency (FFG Forschungsförderungsgesellschaft), project 845020, 841150 and project 848876.

Acknowledgments: The X-ray center (XRC) of TU Wien is kindly acknowledged for the access to the powder X-ray diffractometer.

Conflicts of Interest: The authors declare no conflict of interest. The funders had no role in the design of the study; in the collection, analyses, or interpretation of data; in the writing of the manuscript, and in the decision to publish the results.

References

1. Rahm, D. *Sustainable Energy and the States, Essay on Politics Markets and Leadership*, 1st ed.; McFarland: Jefferson, NC, USA, 2002.
2. IEA. Heating without Global Warming: Market Developments and Policy Considerations for Renewable Heat. Available online: https://www.google.com.hk/url?sa=t&rct=j&q=&esrc=s&source=web&cd=1&cad=rja&uact=8&ved=2ahUKEwiWq76i5-ndAhULUd4KHYqlAxoQFjAAegQICRAC&url=https%3A%2F%2Fwww.iea.org%2Fpublications%2Ffreepublications%2Fpublication%2FFeaturedInsight_HeatingWithoutGlobalWarming_FINAL.pdf&usg=AOvVaw1PEQh4uZihy8UNUEKlI6EJ (accessed on 31 August 2018).

3. IEA. Co-Generation and Renewables. Solutions for a Low-Carbon Energy Future. Available online: https://www.iea.org/publications/freepublications/publication/co-generation-and-renewables-solutions-for-a-low-carbon-energy-future.html (accessed on 31 August 2018).
4. Bauer, T.; Steinmann, W.-D.; Laing, D.; Tamme, R. Thermal energy storage materials and systems. *Annu. Rev. Heat Transf.* **2012**, *15*, 131–177. [CrossRef]
5. Hasnain, S.M. Review on sustainable thermal energy storage technologies, part i: Heat storage materials and techniques. *Energy Convers. Manag.* **1998**, *39*, 1127–1138. [CrossRef]
6. Xu, J.; Wang, R.Z.; Li, Y. A review of available technologies for seasonal thermal energy storage. *Sol. Energy* **2014**, *103*, 610–638. [CrossRef]
7. Zhang, H.; Baeyens, J.; Cáceres, G.; Degrève, J.; Lv, Y. Thermal energy storage: Recent developments and practical aspects. *Prog. Energy Combust. Sci.* **2016**, *53*, 1–40. [CrossRef]
8. Cabeza, L.F.; Castell, A.; Barreneche, C.; de Gracia, A.; Fernández, A.I. Materials used as PCM in thermal energy storage in buildings: A review. *Renew. Sustain. Energy Rev.* **2011**, *15*, 1675–1695. [CrossRef]
9. Dinker, A.; Agarwal, M.; Agarwal, G.D. Heat storage materials, geometry and applications: A review. *J. Energy Inst.* **2015**, *90*, 1–11. [CrossRef]
10. Abedin, A.H.; Rosen, M.R. A critical review of thermochemical energy storage systems. *Open Renew. Energy J.* **2011**, *4*, 42–46. [CrossRef]
11. Cot-Gores, J.; Castell, A.; Cabeza, L.F. Thermochemical energy storage and conversion: A-state-of-the-art review of the experimental research under practical conditions. *Renew. Sustain. Energy Rev.* **2012**, *16*, 5207–5224. [CrossRef]
12. Solé, A.; Fontanet, X.; Barreneche, C.; Martorell, I.; Fernández, A.I.; Cabeza, L.F. Parameters to take into account when developing a new thermochemical energy storage system. *Energy Procedia* **2012**, *30*, 380–387. [CrossRef]
13. Yan, T.; Wang, R.Z.; Li, T.X.; Wang, L.W.; Fred, I.T. A review of promising candidate reactions for chemical heat storage. *Renew. Sustain. Energy Rev.* **2015**, *43*, 13–31. [CrossRef]
14. Pan, Z.; Zhao, C.Y. Dehydration/hydration of mgo/h2o chemical thermal storage system. *Energy* **2015**, *82*, 611–618. [CrossRef]
15. Kuleci, H.; Schmidt, C.; Rybacki, E.; Petrishcheva, E.; Abart, R. Hydration of periclase at 350 °C to 620 °C and 200 mpa: Experimental calibration of reaction rate. *Mineral. Petrol.* **2015**, *110*, 1–10. [CrossRef]
16. Myagmarjav, O.; Ryu, J.; Kato, Y. Dehydration kinetic study of a chemical heat storage material with lithium bromide for a magnesium oxide/water chemical heat pump. *Prog. Nucl. Energy* **2015**, *82*, 153–158. [CrossRef]
17. Birchal, V.S.; Rocha, S.D.F.; Mansur, M.B.; Ciminelli, V.S.T. A simplified mechanistic analysis of the hydration of magnesia. *Can. J. Chem. Eng.* **2001**, *79*, 507–511. [CrossRef]
18. Zhou, S. Hydration Mechanisms of Magnesia-Based Refractory Bricks. Ph.D. Thesis, The University of Brithish Columbia, Vancouver, BC, Canada, December 2004.
19. Ishitobi, H.; Hirao, N.; Ryu, J.; Kato, Y. Evaluation of heat output densities of lithium chloride-modified magnesium hydroxide for thermochemical energy storage. *Ind. Eng. Chem. Res.* **2013**, *52*, 5321–5325. [CrossRef]
20. Myagmarjav, O.; Ryu, J.; Kato, Y. Lithium bromide-mediated reaction performance enhancement of a chemical heat-storage material for magnesium oxide/water chemical heat pumps. *Appl. Therm. Eng.* **2014**, *63*, 170–176. [CrossRef]
21. Shkatulov, A.; Ryu, J.; Kato, Y.; Aristov, Y. Composite material "$Mg(OH)_2$/vermiculite": A promising new candidate for storage of middle temperature heat. *Energy* **2012**, *44*, 1028–1034. [CrossRef]
22. Zamengo, M.; Ryu, J.; Kato, Y. Composite block of magnesium hydroxide – expanded graphite for chemical heat storage and heat pump. *Appl. Therm. Eng.* **2014**, *69*, 29–38. [CrossRef]
23. Müller, D.; Knoll, C.; Ruh, T.; Artner, W.; Welch, J.M.; Peterlik, H.; Eitenberger, E.; Friedbacher, G.; Harasek, M.; Blaha, P.; et al. Calcium doping facilitates water dissociation in magnesium oxide. *Adv. Sustain. Syst.* **2018**, *2*, 1700096. [CrossRef]
24. Kato, Y.; Sasaki, Y.; Yoshizawa, Y. Magnesium oxide/water chemical heat pump to enhance energy utilization of a cogeneration system. *Energy* **2005**, *30*, 2144–2155. [CrossRef]
25. Morozov, S.A. Synthesis of porous magnesium oxide by thermal decomposition of basic magnesium carbonate. *Russ. J. Gen. Chem.* **2003**, *73*, 37–42. [CrossRef]

26. Shand, M.A. *The Chemistry and Technology of Magnesia*; John Wiley & Sons Inc.: Hoboken, NJ, USA, 2006. Available online: https://www.wiley.com/en-us/The+Chemistry+and+Technology+of+Magnesia-p-9780471656036 (accessed on 31 August 2018).
27. Criado, Y.A.; Alonso, M.; Abanades, J.C. Kinetics of the CaO/Ca(OH)$_2$ hydration/dehydration reaction for thermochemical energy storage applications. *Ind. Eng. Chem. Res.* **2014**, *53*, 12594–12601. [CrossRef]
28. Müller, D.; Knoll, C.; Artner, W.; Welch, J.M.; Freiberger, N.; Nilica, R.; Eitenberger, E.; Friedbacher, G.; Harasek, M.; Hradil, K.; et al. Influence of the particle morphology on cycle stability and hydration behavior of magnesium oxide. *Appl. Energy* **2017**, submitted.
29. Hu, X.L.; Carrasco, J.; Klimeš, J.; Michaelides, A. Trends in water monomer adsorption and dissociation on flat insulating surfaces. *Phys. Chem. Chem. Phys.* **2011**, *13*, 12447–12453. [CrossRef] [PubMed]
30. Stankic, S.; Bernardi, J.; Diwald, O.; Knözinger, E. Optical surface properties and morphology of Mgo and Cao nanocrystals. *J. Phys. Chem. B* **2006**, *110*, 13866–13871. [CrossRef] [PubMed]
31. Kazimirov, V.Y.; Smirnov, M.B.; Bourgeois, L.; Guerlou-Demourgues, L.; Servant, L.; Balagurov, A.M.; Natkaniec, I.; Khasanova, N.R.; Antipov, E.V. Atomic structure and lattice dynamics of Ni and Mg hydroxides. *Solid State Ion.* **2010**, *181*, 1764–1770. [CrossRef]
32. Effenberger, H.; Mereiter, K.; Zemann, J. Crystal structure refinements of magnesite, calcite, rhodochrosite, siderite, smithonite, and dolomite, with discussion of some aspects of the stereochemistry of calcite type carbonates. *Zeitschrift für Kristallographie–Cryst. Mater.* **1981**, *156*. [CrossRef]
33. Chen, X.-A.; Song, F.-P.; Chang, X.-A.; Zang, H.-G.; Xiao, W.-Q. A new polymorph of magnesium oxalate dihydrate. *Acta Crystallogr. Sect. E Struct. Rep. Online* **2008**, *64*, m863. [CrossRef] [PubMed]
34. Boiocchi, M.; Caucia, F.; Merli, M.; Prella, D.; Ungaretti, L. Crystal-chemical reasons for the immiscibility of periclase and wüstite under lithospheric p,t conditions. *Eur. J. Mineral.* **2001**, *13*, 871–881. [CrossRef]
35. Brunauer, S.; Emmett, P.H.; Teller, E. Adsorption of gases in multimolecular layers. *J. Am. Chem. Soc.* **1938**, *60*, 309–319. [CrossRef]
36. Degen, T.; Sadki, M.; Bron, E.; König, U.; Nénert, G. The highscore suite. *Powder Diffr.* **2014**, *29*, S13–S18. [CrossRef]
37. Beaucage, G. Approximations leading to a unified exponential/power-law approach to small-angle scattering. *J. Appl. Crystallogr.* **1995**, *28*, 717–728. [CrossRef]
38. Kinning, D.J.; Thomas, E.L. Hard-sphere interactions between spherical domains in diblock copolymers. *Macromolecules* **1984**, *17*, 1712–1718. [CrossRef]
39. Pabisch, S.; Feichtenschlager, B.; Kickelbick, G.; Peterlik, H. Effect of interparticle interactions on size determination of zirconia and silica based systems—A comparison of SAXS, DLS, BET, XRD and TEM. *Chem. Phys. Lett.* **2012**, *521*, 91–97. [CrossRef] [PubMed]

© 2018 by the authors. Licensee MDPI, Basel, Switzerland. This article is an open access article distributed under the terms and conditions of the Creative Commons Attribution (CC BY) license (http://creativecommons.org/licenses/by/4.0/).

Article

Synthesis of Me Doped Mg(OH)$_2$ Materials for Thermochemical Heat Storage

Elpida Piperopoulos [1,2,*], **Marianna Fazio** [1] **and Emanuela Mastronardo** [3,4]

1. Department of Engineering, University of Messina, 98166 Messina, Italy; faziom@unime.it
2. National Interuniversity Consortium of Materials Science and Technology (INSTM), 50121 Firenze, Italy
3. Institute of Advanced Studies of Madrid (IMDEA), Thermochemical Processes Unit, 28935 Madrid, Spain; emanuela.mastronardo@northwestern.edu
4. Department of Materials Science and Engineering, Northwestern University, Evanston, IL 60208, USA
* Correspondence: epiperopoulos@unime.com; Tel.: +39-090-3977-558

Received: 13 June 2018; Accepted: 19 July 2018; Published: 26 July 2018

Abstract: In order to investigate the influence of metal (Me) doping in Mg(OH)$_2$ synthesis on its thermochemical behavior, Ca^{2+}, Co^{2+} and Ni^{2+} ions were inserted in Mg(OH)$_2$ matrix and the resulting materials were investigated for structural, morphological and thermochemical characterization. The densification of the material accompanied by the loss in porosity significantly influenced the hydration process, diminishing the conversion percentage and the kinetics. On the other hand, it increased the volumetric stored/released heat capacity (between 400 and 725 MJ/m^3), reaching almost three times the un-doped Mg(OH)$_2$ value.

Keywords: magnesium hydroxide; thermochemical heat storage; metal doping

1. Introduction

The Renewable Energy Directive establishes an overall policy for the production and promotion of energy from renewable sources in the European Union (EU). The EU target for 2020 is to achieve at least 20% of its total energy requests with renewables. EU countries have already agreed on a new renewable energy target of at least 27% as climate goals for 2030. On 30 November 2016, the European Commission published a proposal for a revised Renewable Energy Directive to make the EU a global leader in renewable energy. Renewable energy can be produced from a wide variety of sources including solar, wind, hydro, geothermal, tidal and biomass. By using more renewables to meet its energy needs, the EU lowers its dependence on imported fossil fuels and makes its energy production more sustainable. Due to climatic variability, the means of storing these types of renewable energy have become an urgent consideration [1]. This has led to the search for efficient and sustainable methods of storing energy and a considerable effort to understand how energy storage works, how existing methods can be improved and how new ones can be developed. Thermal energy storage (TES) transfers heat to storage media during the charging period and releases it at a later stage during the discharging step. It can be usefully applied in solar plants, or in industrial processes. Through TES systems, heat can be stored in the form of sensible [2] or latent heat [3] or in the form of chemical energy (thermochemical storage) [4]. Sensible heat storage is achieved by varying the temperature of a storage material. Latent heat storage is realized changing a material phase at a constant temperature, while the thermochemical storage promotes a reversible chemical reaction. Sensible heat storage is well-documented. Latent heat storage, using phase change materials (PCMs), has been heavily researched and is widely used domestically and industrially. Thermochemical heat storage (TCS) is still at an early stage of laboratory and pilot research in spite of its attractive application for long-term energy storage and higher stored/released heat values [5,6]. Storage density, in terms of the amount of energy per unit of volume, is important for optimizing the use of these kind of materials [7] as it is relevant to their transportation and application in concentrated systems [6]. In 1978, Bowery et al. [8] investigated the practical feasibility of a BaO_2/BaO system for high-temperature heat storage. Theoretical calculations discovered that the endothermic reaction occurred when the temperature exceeds 754 °C, and the calculated energy storage density was about 2.9 GJ/m^3. Subsequently, the reaction was found difficult to achieve complete conversion; even if the temperature rose to 1027 °C, the theoretical conversion rate of BaO_2 had a maximum of 85%. Since then, several TCS materials have been studied and many strategies have been adopted to improve these storage materials [9]. Carrillo et al. studied the effect that co-doping of Mn oxides with Fe and Cu has on the redox temperatures of both forward and reverse reactions [10,11]. Block et al. tested several compositions of eight binary metal oxide systems as well as the pure metal oxides (cobalt oxide, iron oxide, copper oxide and manganese oxide) in terms of their ability to store energy thermochemically [12]. The calcium oxide hydration/dehydration reaction is proposed as a suitable reaction couple for thermochemical energy storage systems for its high energy density (0.4 kWh/kg) and low material cost (50 €/t) [13–15]. Sakellariou et al. prepared mixed calcium oxide–alumina compositions, assessed in terms of their cyclic hydration–dehydration performance in the temperature range of 200–550 °C. One of the main purposes of using Al as additive was related to materials structural enhancement [16]. A suitable TCS system storing in lower temperature range between 200 °C and 400 °C, which has been examined in this study, is the dehydration/hydration reaction of magnesium hydroxide/oxide:

$$Mg(OH)_2(s) \leftrightarrow MgO(s) + H_2O(v) \quad \Delta H_0 = \pm 81 \text{ kJ/mol} \quad (1)$$

The above system offers several advantages, high storage capacity, medium operating temperature range (as reported above), long-term storage of reactants and products, low heat loss and non-toxicity of the materials [17]. Through the endothermic dehydration reaction, heat can be stored and released when required by the reverse exothermic hydration reaction. This system has been widely studied to improve storage material performances, as mass and volume energy density, kinetics and ciclability. Shkatulov et al. studied $LiNO_3$-doped $Mg(OH)_2$ storage material that exhibits a decrease in the dehydration start temperature by 76 °C [18]. Junichi et al. developed a 6.8 wt.% $LiCl/Mg(OH)_2$ system that drops the dehydration temperature of magnesium hydroxide, from 277 °C to 233 °C, being able to store 816 MJ/m^3 volumetric heat storage capacity [19]. Muller et al. found that calcium doping of magnesium oxide results in significantly increased water dissociation rates, thus enhancing both hydration rate and reaction completeness of hydration compared to pure MgO [20]. Zamengo et al. prepared a $Mg(OH)_2/MgO$ system supported on expanded graphite. The pelletized storage material, decreasing the tablets volume required to store the same amount of thermal energy of $Mg(OH)_2$ pellets of almost 13.6%, increases volume energy density [21]. In previous studies it was found that, synthesizing $Mg(OH)_2$ in presence of a cationic surfactant (cetyl trimethyl ammonium bromide—CTAB), an optimum CTAB concentration exists and it exhibits the highest volumetric stored/released heat capacity, ~560 MJ/m^3 two times higher than that measured over $Mg(OH)_2$ prepared in absence of CTAB [22]. The purpose of this work is to investigate the influence of metal (Ca^{2+}, Co^{2+} and Ni^{2+}) doping in $Mg(OH)_2$ synthesis on its structural and morphological properties and consequently on its thermochemical behavior.

2. Materials and Methods

2.1. Samples Preparation

The Metal (Me) doped $Mg(OH)_2$ samples were synthesized by precipitation method. The following raw materials were used: $Mg(NO_3)_2 \cdot 6H_2O$, 99%, supplied by Sigma-Aldrich (St. Louis, MO, USA), as magnesium source, ammonia solution (NH_4OH, 30 wt.% Carlo Erba) as precipitating agent and $Ca(NO_3)_2$, $Co(NO_3)_2$ and $Ni(NO_3)_2$ respectively for Ca^{2+}, Co^{2+} and Ni^{2+} doping metals. The precipitation was carried out as follows: 50 mL of a solution containing Mg^{2+} and Me ion (Ca^{2+} or Co^{2+} or Ni^{2+}) were gradually added (2.5 mL/min) through a peristaltic pump to 150 mL of NH_4OH

solution (ph = 11.8), under magnetic stirring. The final solution was aged at ambient temperature for 24 h, then it was vacuum filtered (0.22 µm); the collected solid was washed with deionized water and dried in a vacuum oven (Binder, Tuttlingen, Germany) at 50 °C overnight. Table 1 reports the code of samples and the chemical composition of solutions for all the preparations.

Table 1. Sample code, chemical compositions of the solutions. Mg^{2+} and OH^- molar concentration were 0.01 M and 0.063 M in each preparation.

Sample Code	Type of Me^{2+}	[Me]	Me/Mg^{2+} Nominal Molar Ratio
MH	-	-	-
MH-Ca1	Ca^{2+}	0.0003	0.033
MH-Ca2	Ca^{2+}	0.0007	0.067
MH-Ca3	Ca^{2+}	0.0020	0.200
MH-Ni1	Ni^{2+}	0.0003	0.033
MH-Ni2	Ni^{2+}	0.0007	0.067
MH-Ni3	Ni^{2+}	0.0020	0.200
MH-Co1	Co^{2+}	0.0003	0.033
MH-Co2	Co^{2+}	0.0007	0.067
MH-Co3	Co^{2+}	0.0020	0.200

2.2. Samples Characterization

Quantitative analysis of calcium, nickel and cobalt present into the solid was performed by means of ICP-MS spectrometer (PERKIN-ELMER, model NexION 300×, Waltham, MA, US). Approximately 3 mg ($wt_{measured}$) of each synthesized sample ($wt_{synthesized}$) was dissolved in the minimum volume of concentrated HNO_3, and then deionized water was added until the final volume of 10 mL (V_f) was reached. Exactly 100 µL (V_1) of this solution, mixed of 100 µL of concentrated HNO_3, were diluted up to 10 mL (V_2) and then analyzed. The grams of dopant (Me) present in the samples are calculated as follows:

$$Me\ (g) = \{(\frac{[Me]_{ICPMS}\ (ppm) \times V_2(l)}{V_1\ (l)}) \times \frac{V_f(l)}{1000}\} \times \frac{wt_{synthesized}\ (g)}{wt_{measured}\ (g)} \quad (2)$$

Pore volume was calculated by Barrett-Joyner-Halenda (BJH) method using the nitrogen desorption isotherm measured at −196 °C with a Quantachrome Autosorb-iQ MP (NOVA 1200, Boynton Beach, FL, USA) instrument. Samples were degassed prior to analysis under vacuum at 120 °C for 3 h. Each sample's mean particle size was determined by Dynamic Light Scattering (DLS) technique. DLS was measured at 25 °C using a Zetasizer Nano ZS instrument (Malvern Instruments, Malvern, UK) equipped with a helium-neon 4 mW laser (wavelength λ_0 = 632.8 nm). The scattering angle was equal to 173°. Prior to measurements, samples were sonicated for 30 minutes in ethylene glycol. The bulk density of samples was measured by weighing a known volume of solids (V (mL)) and calculated by the formula:

$$\rho = m\ (kg)/V(m^3) \quad (3)$$

The as-prepared samples were analyzed by means of scanning electron microscopy (SEM, Quanta 450, FEI, Hillsboro, OR, USA) and X-Ray Diffraction (XRD, Bruker D8 Advance, Bruker, Billerica, MA, USA) to determine their morphology and crystal structure.

SEM analysis were performed on Cr-metallized samples and operating with an accelerating voltage of 10 kV under high vacuum conditions (6.92 × 10^{-5} Pa).

2.3. Thermochemical Performance

The evaluation of the thermochemical behavior of the prepared samples under cyclic heat storage/release experiments was performed using a customized thermogravimetric unit (STA 449 F3 Jupiter Netzsch, Selb, Bavaria, Germany) that allowed us to carry out a succession of dehydration and hydration reactions. The thermogravimetric apparatus was equipped with a water vapor generator for the vapor supply during the hydration reaction. A cyclic heat storage/release experiment was carried out on a mass of ~15 mg as reported elsewhere [17,23,24]: the sample was first dried at 125 °C in inert atmosphere (under N_2 flow: 100 mL/min) for 60 min to remove the physically adsorbed water. Then, the temperature was increased at 10 °C/min up to the dehydration temperature (T_d = 350 °C) and dehydration reaction proceeded over 120 min under isothermal conditions. After the complete dehydration reaction, the temperature was decreased (cooling rate = -10 °C/min) to the hydration temperature (T_h = 125 °C). The hydration reaction proceeded over 120 min, during which the water vapor necessary for the re-hydration reaction was supplied by the water vapor generator at 2.2 g/h and mixed with 35 mL/min N_2 as carrier gas (p_{H2O} = 57.8 kPa). After the fixed hydration time, the water vapor supply was stopped and the sample was kept at 125 °C for 30 min under a constant N_2 flow (100 mL/min) to remove physically adsorbed water from the sample. This procedure was repeated for each heat storage/release cycle. In this study, for a preliminary comparison, the samples were subjected to 3 cycles experiments. To be consistent with previous studies [17,22,23,25] the materials performances were expressed in terms of reacted fraction (β(%)) defined by Equation (4):

$$\beta(\%) = (1 - \frac{\Delta m_{real}}{\Delta m_{th}}) \times 100, \qquad (4)$$

where Δm_{real}(%) was the instantaneous real mass change and Δm_{th}(%) was the theoretical mass change due to the dehydration of 1 mol $Mg(OH)_2$, respectively expressed by Equations (5) and (6):

$$\Delta m_{real}(\%) = \frac{m_{in} - m_{inst}}{m_{in}} \times 100, \qquad (5)$$

$$\Delta m_{th}(\%) = (\frac{M_{Mg(OH)2} - M_{MgO}}{M_{Mg(OH)2}}) \times 100 = 30.89\%, \qquad (6)$$

where m_{in}(g) and m_{inst}(g) were respectively the initial sample mass and the instantaneous mass during TG analysis. While, $M_{Mg(OH)2}$(g/mol) and M_{MgO}(g/mol) were respectively the molecular weight of $Mg(OH)_2$ and MgO.

The dehydration and hydration conversions ($\Delta\beta_{d/h}$(%)) were calculated respectively by Equations (7) and (8):

$$\Delta\beta_d(\%) = \beta_d^i - \beta_d^f, \qquad (7)$$

$$\Delta\beta_h(\%) = \beta_h - \beta_d^f, \qquad (8)$$

where β_d^i and β_d^f were respectively the reacted fraction at the beginning and at the end of the dehydration treatment. While, β_h was the final reacted fraction of MgO at the point of water supply termination.

The stored/released heat capacity per volume unit ($Q_{s/r}^V$ (MJ/m^3)) was calculated using Equation (9):

$$Q_{s/r}^V (MJ/m^3) = -\frac{\Delta H^0}{M_{Mg(OH)2}} \times \Delta\beta_{d/h} \times \rho \qquad (9)$$

where ΔH^0 (kJ/mol) is the enthalpy of reaction and ρ (kg/m^3) the bulk density of the sample.

3. Results and Discussion

3.1. Me Doped Mg(OH)$_2$ Preparation

In the first instance, it was evaluated whether, under the preparation condition of the present work, each ion could precipitate as hydroxide. Precipitation of hydroxide from the solution through the reaction (10)

$$Me^{n+} + nOH^- \rightarrow Me(OH)_n \text{ (s)}, \qquad (10)$$

occurs when the supersaturation conditions are reached. Supersaturation conditions are defined as:

$$[Me^{n+}] \cdot [OH^-]^n > K_{sp}, \quad (11)$$

where $[Me^{n+}]$ and $[OH^-]$ represent the concentration expressed as molarity (M) of cation and hydroxyl ions, n represent the hydroxyl's stoichiometric coefficient and K_{sp} is the thermodynamic equilibrium constant of solubility product. As shown in Table 2 supersaturation conditions are satisfied in case of $Mg(OH)_2$, $Co(OH)_2$ and $Ni(OH)_2$ formation but not for $Ca(OH)_2$, whatever the calcium concentration used being the ionic product $[Me^{n+}] \cdot [OH^-]^n < K_{sp}$.

Table 2. Evaluation of supersaturation conditions for $Mg(OH)_2$, $Co(OH)_2$, $Ni(OH)_2$ and $Ca(OH)_2$ formation under conditions used in the present work. Y: Yes, N: No.

Hydroxide	$[Me^{2+}] [OH^-]^n$	K_{sp} (at 25 °C)	Supersaturation Condition
$Mg(OH)_2$	3.98×10^{-7}	1.80×10^{-11}	Y
$Ca(OH)_2$	1.33×10^{-8}	7.90×10^{-06}	N
$Ca(OH)_2$	2.65×10^{-8}	7.90×10^{-06}	N
$Ca(OH)_2$	7.96×10^{-8}	7.90×10^{-06}	N
$Ni(OH)_2$	1.33×10^{-8}	2.80×10^{-16}	Y
$Ni(OH)_2$	2.65×10^{-8}	2.80×10^{-16}	Y
$Ni(OH)_2$	7.96×10^{-8}	2.80×10^{-16}	Y
$Co(OH)_2$	1.33×10^{-8}	2.50×10^{-16}	Y
$Co(OH)_2$	2.65×10^{-8}	2.50×10^{-16}	Y
$Co(OH)_2$	7.96×10^{-8}	2.50×10^{-16}	Y

As will be further explained, in reality neither $Ni(OH)_2$ nor $Co(OH)_2$ solids form (Figure 1). This is due to the fact that with a large excess of ammonia, cobalt and nickel hydroxides redissolve forming hexaminocobalt(II) ($Co(NH_3)_6^{2+}$) and hexaminonickel(II) ($Ni(NH_3)_6^{2+}$) ions as ammonia substitutes as a ligand [25]. As shown in Figure 1 no solid formation occurs even after 24 h. In case of Co^{2+} solution, pink colored due to presence of $Co(H_2O)_6^{2+}$, upon addition of NH_4OH color rapidly changes to yellow then to a deep red-brown. This is due to oxidation of hexaminocobalt(II) to hexaminocobalt(III) ions by air [25]. In case of Ni^{2+}, light green colored by the complex $Ni(H_2O)_6^{2+}$, addition of ammonia causes a color change to light blue typical of $Ni(NH_3)_6^{2+}$ complex [25].

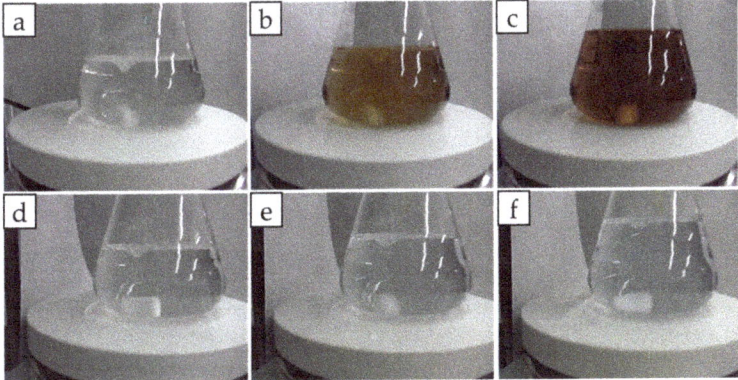

Figure 1. Formation of cobalt and nickel hexamine complexes. Starting aqueous Co^{2+} solution 0.002 M (**a**); Upon addition of $NH_4(OH)$ (**b**) and after mixing for 24 h (**c**). Starting aqueous Ni^{2+} solution 0.002 M (**d**); Upon addition of $NH_4(OH)$ (**e**) and after mixing for 24 h (**f**).

After these preliminary evaluations MH, MH-Ca, MH-Co, and MH-Ni were prepared according to the procedure reported in the experimental section. The Me content on the final sample, $(g_{Me}/g_{sample})\%$, is reported in Figure 2. As general feature, regardless the type of Me, the load increases with the initial amount present into the solution. At given Me initial concentration the g_{Me}/g_{sample} content varies among the different type of Me; in case of Ni the lowest amount of loaded Me is obtained while the highest amount is achieved for Co containing samples.

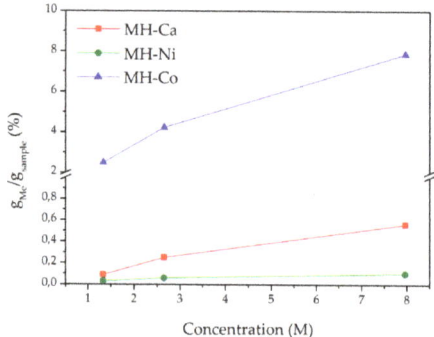

Figure 2. Me content per g of sample obtained by precipitation vs. Me concentration in the starting solution.

Considering that no calcium, cobalt and nickel hydroxide form, it can be assumed that these metal ions are included into $Mg(OH)_2$ host matrix.

3.2. Structure and Morphology of Samples

In Figure 3a–d XRD analysis of MH and MH-Me samples are shown. The diffractograms are acquired in a 2θ range between 10° and 80°. $Mg(OH)_2$ spectrum (Figure 3a) presents the reflection peaks typical of hexagonal brucite (2θ: 18.5°, 32.5°, 38°, 51°, 58.5°, 62°, 68°, 72°), in agreement with standard data, (JCPDS 7-0239 and JCPDS 25-0284). The three most intense peaks (2θ = 18.5°, 38.0°, 58.5°) appear sharp and narrow as a result of high degree of crystallization of hexagonal brucite. Reflection peaks of MH-Ca and MH-Ni samples, regardless the amount of calcium or nickel, match with those of pure $Mg(OH)_2$ in terms of peaks position (Figure 3b,c). They are intense and narrow thus suggesting that the high crystallization degree of brucite is maintained. For MH-Ca2 and MH-Ca3 samples (Figure 3b) is also present a peak at 29.4°, related to $CaCO_3$ (JCPDS 47-1743) likely due to the slight carbonation of calcium ions by CO_2 present in the atmosphere. The main difference with respect to pure MH concerns the change in the relative intensity among the two main peaks relative to (001) and (101) plane. The intensity of $Mg(OH)_2$ (001) plane's peak, which corresponds to the basal plane of brucite, becomes stronger than the diffraction peak for the (101) plane. As reported in Table 3 Entries 1–7, the intensity ratio of reflections I_{001}/I_{101} increases from 0.78 (MH sample) up to values ranging between 0.95–1.34 for MH-Ca e MH-Ni. No clear correlation is observed between the increase of I_{001}/I_{101} and the metal content. From these results it is possible to conclude that in presence of calcium and nickel, ions preferential growth along the (001) hexagonal basal plane of brucite occurs leading to a layered structure, e.g., flakes or platelets, with high aspect ratio along the c-axis. [26–28]. Wu et al. have already reported that the strength of (001) plane became stronger than that of (101) plane upon hydrothermal modification of $Mg(OH)_2$ in presence of $CaCl_2$ [29]. MH-Co samples, instead, shows a peculiar feature. The spectra shown in Figure 3d present reflection peaks, centered at the same position of those of brucite (Figure 3a), with a progressive intensity decrease and peak broadening, as Co content increases, that indicate the lowering of crystallization degree. Rietveld refinement reported in Table 3 confirms that metal ions are included into $Mg(OH)_2$ host matrix because a volume cell ($V(\text{Å}^3)$) increase is observed, in relation to the metal load. At lower metal load for all Me-doped samples the volume cell remains almost similar to the MH sample's one, but increasing Me load it increases till 41.5 Å^3 for MH-Co3 sample (Entry 9 in Table 3), which presents the higher amount of Co in the matrix. Only for MH-Ni2 and MH-Ni3 (Entries 6 and 7 in Table 3) it decreases. Substituting Mg ion (r_{Mg2+} = 0.72 Å) with Ca ion (r_{Ca2+} = 100 Å) it is simple to understand, according to Vegard's law [30], the cell volume change, while it is more difficult in the case of Co (r_{Co2+} = 0.70 Å) and Ni (r_{Ni2+} = 0.70 Å) ions, which ion radius are similar to Mg's one. In these cases, probably, atoms are substituted interstitially leading to a lattice expansion [31].

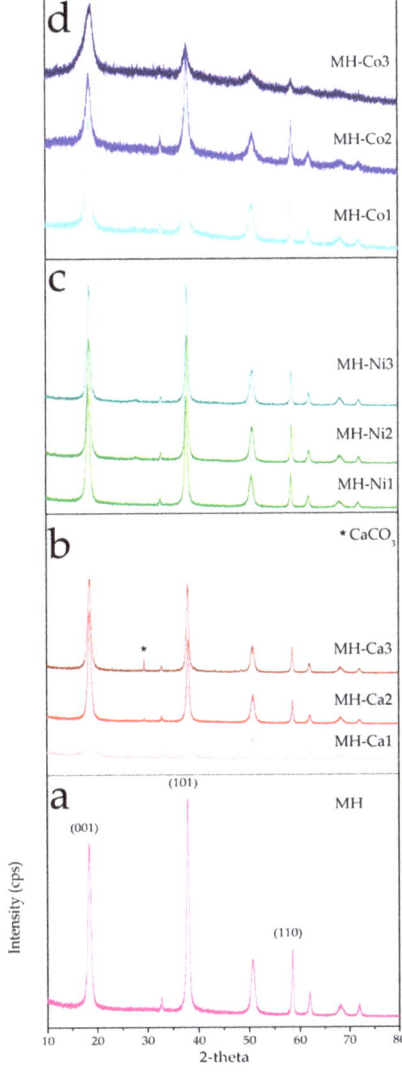

Figure 3. XRD patterns of MH (**a**), MH-Ca (**b**), MH-Ni (**c**) and MH-Co (**d**) samples.

The morphology of MH and MH-Me samples is evaluated by means of SEM analysis, shown in Figure 4a–k. MH sample (Figure 4a) presents as large aggregates prevalently formed by magnesium hydroxide hexagonal platelets, in agreement with XRD findings; in addition, a few rounded shaped particles (red arrows) are also visible. The evolution of the sample morphology as the result of the doping by calcium and nickel appears very similar. In particular, increasing the amount of calcium and nickel in the solid, large agglomerates of highly stacked hexagonal brucite particles form (Figure 4b–g). The use of cobalt as doping ion, instead, gives rise to a dramatic change in the morphology with respect to MH, especially at higher cobalt load. Indeed, while for MH-Co1 brucite platelets having the peculiar stacked configuration are still visible (Figure 4h), in case of MH-Co2 and MH-Co3 it is clearly observed the progressive formation of amorphous hydroxide. In particular, for MH-Co2 sample it can be seen magnesium hydroxide platelets (Figure 4i white arrow) embedded into large portions of badly crystallized material (Figure 4i black arrow). Increasing the cobalt content, MH-Co3 sample, the crystalline hexagonal brucite is practically not visible anymore or it is very difficult to distinguish, and large sheets of poorly crystallized hydroxide represent the material's main component (Figure 4j). SEM analysis is in agreement with the XRD results that evidence the progressive amorphization of $Mg(OH)_2$ increasing the cobalt content (Figure 4h–j).

It is noteworthy from the low magnification images of MH-Co3 sample (Figure 4k) that the large sheets of unshaped badly crystallized hydroxide are very densely packed, forming a continuous and extended rough surface.

Referring to the mechanism of hexagonal $Mg(OH)_2$ growth, based on the model of anion coordination polyhedron (ACP) [32] where the nucleation seeds $Mg(OH)_6^{4-}$ first form the growth units (Figure 5a) that pile up with each other forming large dimension growth units in the same face (x, y) (Figure 5b) which then connect one to another along the z axis forming (001) planes (Figure 5c) and finally the hexagonal structure. It can be argued that calcium and nickel ions promote the growth along the z axis (Figure 5c) then the hexagonal structure, as inferred by the increase of the intensity ratio I_{001}/I_{101} while cobalt ions, instead, strongly hinder the piling of growth units in the x, y plane and then the crystal formation.

The mean particle size of investigated samples, as inferred by DLS analysis, are reported in Table 3.

Table 3. Intensity ratios and morphological properties of investigated samples.

Entry	Sample Code	Intensity Ratios		Rietveld Refinement	Morphological Properties		
		$I_{001/101}$	$I_{001/110}$	$V(Å^3)$	Mean Particle Size (nm) *	ρ (kg/m^3)	V_{pore} (cm^3/g)
1	MH	0.78	2.63	40.9	180.5 ± 24.0	350	0.618
2	MH-Ca1	1.16	3.64	40.7	78.9 ± 44.5	685	0.614
3	MH-Ca2	1.34	4.77	40.8	131.3 ± 33.1	644	0.497
4	MH-Ca3	1.04	3.57	41.3	117.1 ± 45.6	740	0.525
5	MH-Ni1	1.09	3.26	40.8	97.8 ± 50.1	752	0.885
6	MH-Ni2	0.97	3.19	41.3	105.8 ± 38.1	712	0.478
7	MH-Ni3	0.95	3.26	41.0	118.4 ± 20.6	663	0.515
8	MH-Co1	0.89	2.36	40.4	85.8 ± 34.0	616	0.778
9	MH-Co2	0.72	1.69	40.9	67.1 ± 40.0	587	0.914
10	MH-Co3	1.85	4.84	41.5	188.6 ± 23.4	1.050	0.245

* Measured by means of Dynamic Light Scattering analysis.

MH shows the highest value centered at 181 nm. A decrease of the mean particle size is obtained for all MH-Me1 (Entries 2, 5, 8). At higher Me content, two different behaviors have been observed depending on the type of metal. For calcium and nickel doped samples mean particle size returns progressively to increase, although it is lower than that of MH, with the metal content (Entries 3, 4, 6, 7).

For MH-Co2 instead, mean particle size continues to decrease in MH-Co2 (Entry 9) while abruptly increases for MH-Co3 sample, containing the highest metal content (Entry 10).

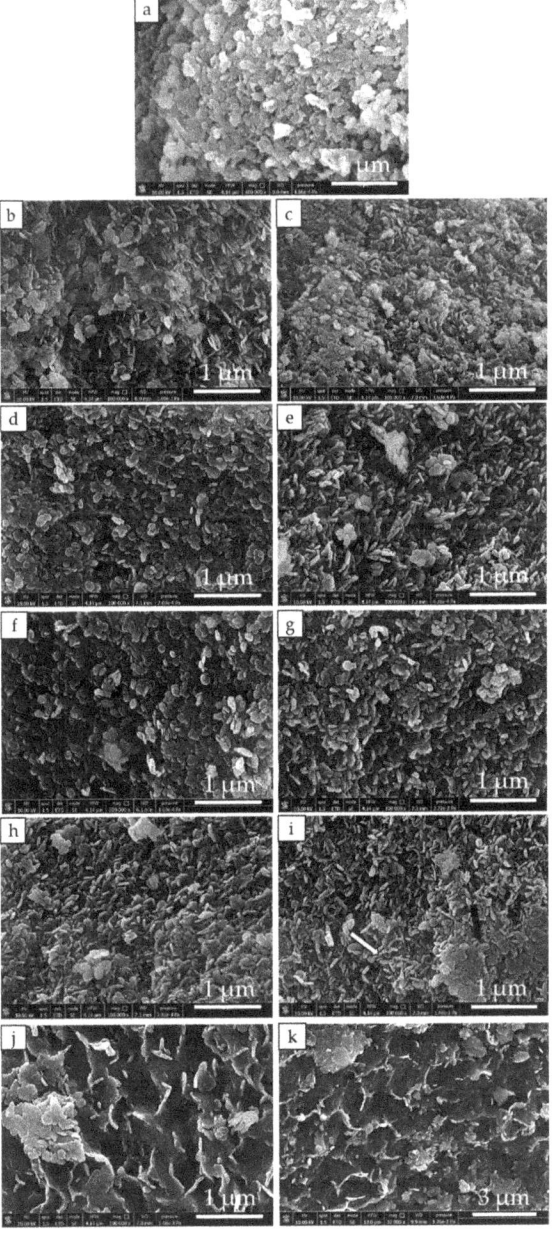

Figure 4. SEM images of investigated samples. MH (**a**); MH-Ca1 (**b**); MH-Ca2 (**c**); MH-Ca3 (**d**); MH-Ni1 (**e**); MH-Ni2 (**f**); MH-Ni3 (**g**); MH-Co1 (**h**); MH-Co2 (**i**); MH-Co3 (**j**,**k**).

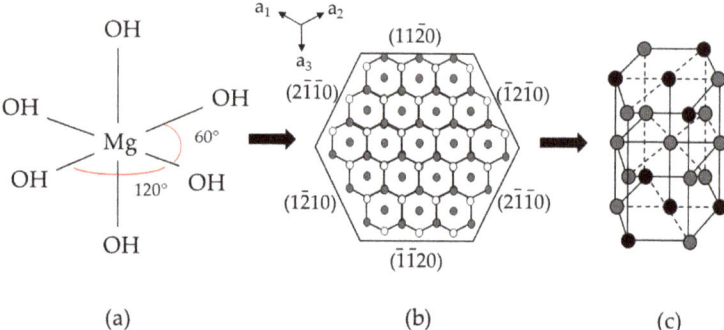

Figure 5. Mg(OH)$_2$ growth, based on the model of anion coordination polyhedron (ACP). Growth unit (**a**); Large dimension growth units in the same face (**b**); Hexagonal structure (**c**).

It is noteworthy that for MH-Ca and MH-Ni samples DLS data really reflects the size of the crystalline hexagonal platelets which seems to be influenced by the cations content. Wu et al. have already reported an increase of Mg(OH)$_2$ particle size increasing the calcium content during hydrothermal treatment of hydroxide [29]. The authors suggest that calcium promotes the formation of Mg(OH)$^+$ which may be favorable for the formation of nucleation seeds Mg(OH)$_6^{4-}$ which represents the growth unit for Mg(OH)$_2$ growth [32]. A similar effect can be depicted for nickel ion.

In the case of MH-Co2 and MH-Co3 samples instead, DLS analysis likely reflects the size of crystalline hexagonal platelets (very few, especially in the case of MH-Co3) in addition to the size of the amorphous portions of the sample. Considering that the preparation of the samples for DLS analysis provides that they are sonicated, it is likely that, due to the lowest mechanical resistance, amorphous phase is fragmented in an uncontrolled way. Therefore, the resulting particle size observed is not the direct evidence of a such influence of cobalt ion during the Mg(OH)$_2$ growth.

Table 3 also lists the material's properties such as apparent density ρ (kg/m^3). From the reported data it is evident that the apparent density of all MH-Me samples is significantly higher than that of pure MH. In general, the apparent density enhancement ranges between 68% (sample MH-Co2) up to 200% (sample MH-Co3). The strong enhancement of density is visually demonstrated in Figure 6.

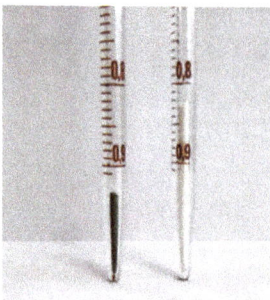

Figure 6. Volume occupied by ~69 mg of MH-Co3 (**on the left**) and MH (**on the right**) samples.

Apparent density is defined as the average density of the material and includes the volume of pores within the particle boundary [33]. Generally, the higher the density, the smaller the pore volume in the sample. The almost general behavior of doped samples (MH-Ca and MH-Ni), in fact, reflects a higher density of the material and a lower value of the porosity, except for the MH-Ni1 sample, which morphology (Figure 4e) appears to be less stacked than the samples with the highest metal load and more similar to MH-Co1 and MH-Co2 (Figure 4h,i), which show a comparable pore volume (Entries 8 and 9 in Table 3). The same peculiar morphology was found for $Mg(OH)_2$ prepared in the presence of CTAB, which promotes the formation of well separated $Mg(OH)_2$ particles, lowering the hydroxide mean particle diameter and increasing the bulk density likely due to the peculiar stacked configuration of hydroxide particles, reported elsewhere [22].

The increase of apparent density could be due to two concomitant effects, which are the lowering of particle size and the strong agglomeration of magnesium hydroxide particles (MH-Ca and MH-Ni sample) or, as in case of MH-Co, to the densely packed amorphous material formation, as evidenced by SEM and by the lowest value of pore volume (0.245 cm^3/g) detected for MH-Co3 sample.

3.3. Thermochemical Behavior

The thermogravimetric data are calculated assuming the metal doping negligible and $Mg(OH)_2$ at 100 wt.%. The curves in Figure 7 are relative to the third cycle, when the thermochemical behavior of the samples was observed to be stable [24].

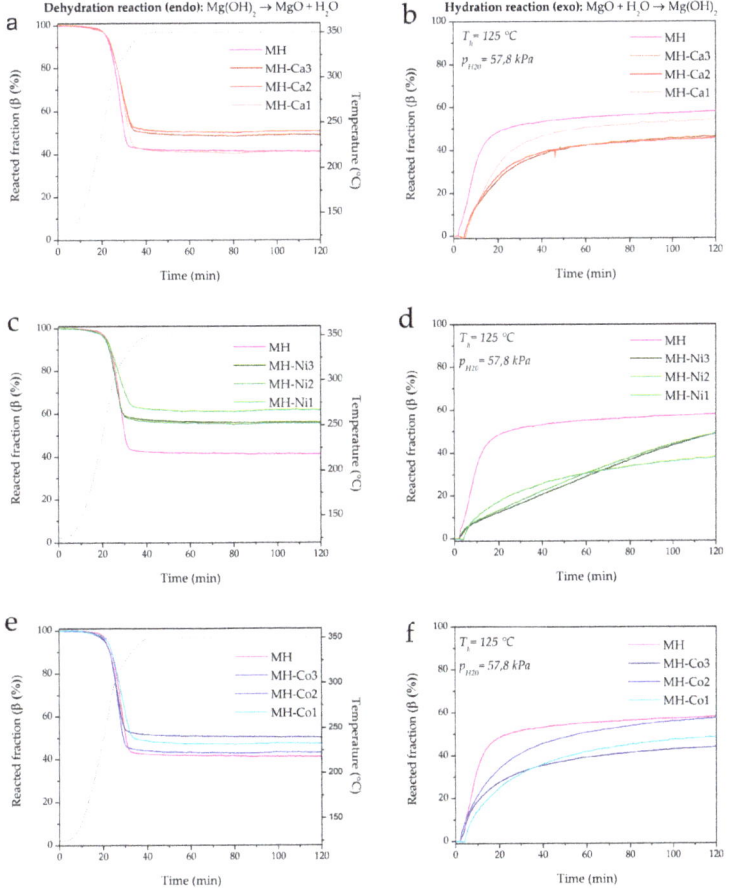

Figure 7. TG analysis, influence of metal loading. Reacted fraction in dehydration and hydration reactions of MH-Ca (**a**,**b**), MH-Ni (**c**,**d**), MH-Co (**e**,**f**).

For all the doped materials, the percentage of MH reacted fraction during dehydration and hydration decreases (Figure 7). MH-Ca and MH-Ni follow the opposite trend observable for morphological properties in Table 3. In fact, Mg(OH)$_2$ conversion progressively decreases, increasing metal load from MH-Ca1 to MH-Ca2 (Entries 2 and 3 in Table 4), and then it remains almost stable for MH-Ca3 (Entry 4 in Table 4). The same behavior is observed for MH-Ni. The mean particle size, as described before, progressively increases following the same criteria. For MH-Co1, instead, conversion (%) continues to increase to MH-Co2 (Entry 9 in Table 4) while abruptly decreases for MH-Co3 sample, which presents the highest mean particle size (Entry 10 in Table 3). Additionally, for hydration, a similar trend is observed. If the 1st cycle dehydration reaction is analyzed, it can be observed that the conversion percentage of all the samples is equal to MH conversion and in some cases also higher (Entries 2, 3, 5, 6 and 7 in Table 4) or quite low (Entries 4 and 10 in Table 4). Therefore, the limiting process that influences the material behavior is the hydration. The β_h%, during first cycle, as shown in Table 4, does not reach MH hydration with the exception of MH-Ca1 (Entry 2 in Table 4), which also maintains the higher conversion percentage in the following cycles. This behavior seems to be related to the main particle size reported in Table 3. As discussed above, the smaller particle size is strictly correlated with the higher density of the doped samples. This morphology strongly influences the magnesia hydration. As reported by Tang et al. [34], MgO hydration process follows common MgO dissolution/Mg(OH)$_2$ precipitation mechanism, well accepted in the literature [35–37]. Initially water vapor is chemisorbed on the MgO and then physically adsorbed to form a liquid layer on the surface of the solid (chemical control of the reaction). This layer of water reacts with the MgO to form a surface layer of Mg(OH)$_2$, that covers surfaces and pores of MgO particles. As a result, the diffusion of water vapor is hindered inside the particles, which reduces the overall reaction rate and the rehydration conversion β_h% (diffusion controlled). When density is high, because of the small particle size and the packed morphology described in Figure 4, the porosity of the material is very poor and the water permeability is difficult. Figure 8 shows the SEM analysis of the investigated samples after cycling. For brevity, only MH and doped samples with highest metal load are reported (MH-Ca3, MH-Ni3, MH-Co3). It can be observed that coalescence is more favored for doped samples. For MH-Ca3 and MH-Ni3, the particle size increase is clearly observable (compare Figure 4 (white arrows) with Figure 8 (white circles)), MH-Co3 keeps its packed structure, formed by large sheets of poorly crystallized hydroxide (red arrows in Figure 8). Probably, also in this case, the high density plays a very important role influencing the change of morphology during the dehydration/hydration cycles. The presence of a lower porosity of the material and a smaller particle size favors the coalescence of the latter in larger particles; this decreases heat transfer property and leads to further loss of bulk porosity diminishing the MgO rehydration kinetics [24].

Figure 8. *Cont.*

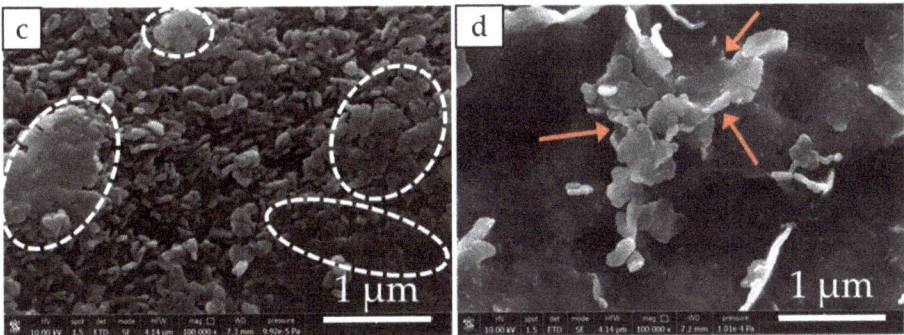

Figure 8. SEM images of investigated samples after cycling: MH (**a**); MH-Ca3 (**b**); MH-Ni3 (**c**); MH-Co3 (**d**).

Also noteworthy is the fact that, from the slope of the dehydration and hydration curves, doped samples exhibit similar dehydration kinetics with respect to un-doped MH (see Figure 7a,c,e). On the contrary, hydration kinetics is highly affected by the doping, which, in general, decreases the kinetics (see Figure 7b,d,f). This is evident especially in Ni-doped samples (see Figure 7d). Hence, depending on the final application of the storage technology, heat can be released at a required rate by tuning MH with a proper dopant cation and amount.

Table 4. Comparison between dehydration/hydration conversions ($\Delta\beta_{d/h}$) at first and third cycles.

Entry	Sample Code	1st Cycle		3rd Cyle		Q_s^V (MJ/m^3)	Q_s^V (MJ/m^3)
		β_d (%)	β_h (%)	β_d (%)	β_h (%)		
1	MH	89.0	61.6	58.8	58.0	285.4	281.95
2	MH-Ca1	90.0	62.7	58.8	54.0	536.4	513.98
3	MH-Ca2	90.0	52.7	49.4	45.6	444.9	408.44
4	MH-Ca3	88.3	55.8	51.2	45.6	525.7	475.5
5	MH-Ni1	93.7	36.6	38.2	38.0	401.7	397.01
6	MH-Ni2	97.7	42.0	44.4	49.0	441.0	485.01
7	MH-Ni3	94.8	41.1	44.4	49.0	407.0	449.36
8	MH-Co1	89.0	55.5	52.5	48.6	450.6	416.22
9	MH-Co2	89.0	56.6	56.8	57.2	463.3	466.98
10	MH-Co3	87.2	55.5	49.7	43.8	724.8	640.75

Looking at stored/released heat capacity by unit volume ($Q_{s/r}^V$) (Figure 9a–f) it can be seen that MH shows the lowest stored/released heat capacity that is 285 MJ/m^3 with respect to MH-M samples for which a higher $Q_{s/r}^V$ is generally observed, as a consequence of the higher apparent density (Table 3).

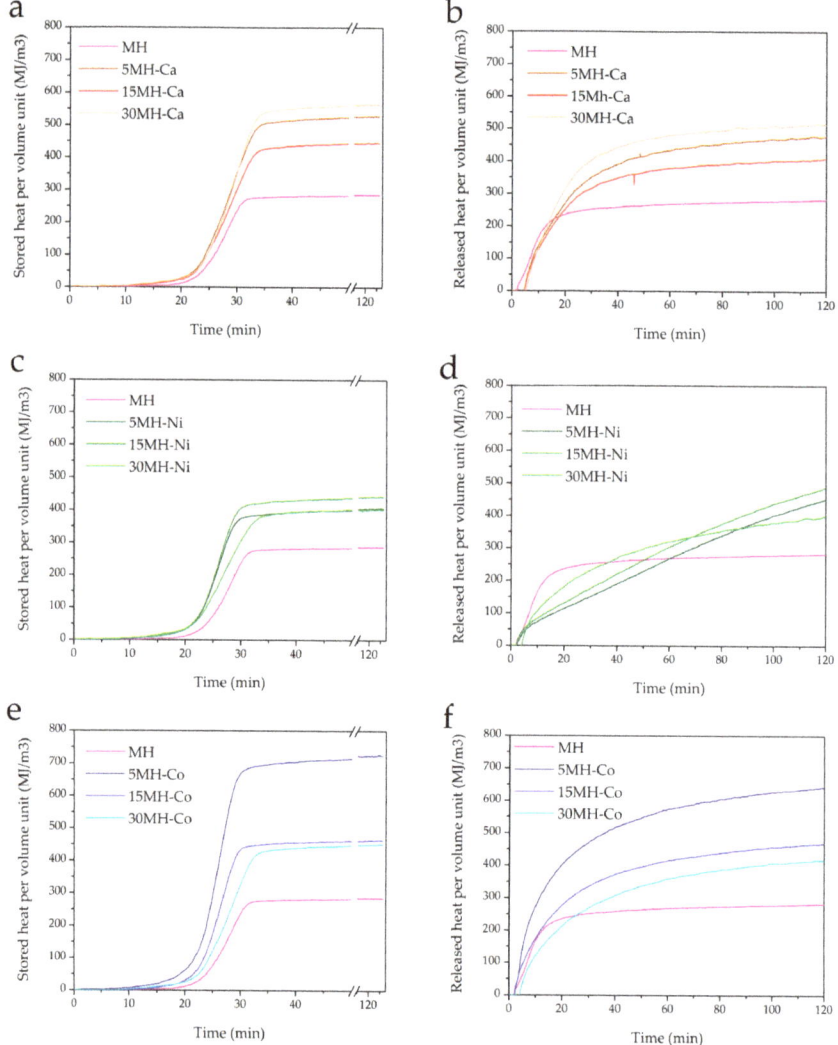

Figure 9. Stored and released heat per volume unit of MH-Ca (**a**,**b**), MH-Ni (**c**,**d**), MH-Co (**e**,**f**).

The highest value 725 MJ/m^3 is achieved on MH-Co3 (Figure 9e), which is almost three times higher than MH's value. This value is, so far, the highest reported in the literature for pure Mg(OH)$_2$ heat storage material. The stored heat increases with an increase in the metal load doping. Released heat per volume unit in doped samples almost never reaches the 100% of stored heat.

4. Conclusions

The present study clearly suggests that morphological characteristics (porosity, mean particle size) and apparent density are significantly influenced by the Me (Ca^{2+}, Ni^{2+}, Co^{2+}) doping during the preparation of Mg(OH)$_2$ through precipitation. It was found that, considering that no calcium, cobalt and nickel hydroxides precipitate during the synthesis, these metal ions are included into Mg(OH)$_2$ host matrix, as confirmed by Rietveld refinement of XRD analysis. All the investigated samples show an apparent density increase. MH-Co3, which presents badly crystallized and highly packed hydroxide, reaches a higher density than the MH sample of 200%. Apparent density describes two concomitant effects that are the lowering of particle size and the strong agglomeration of magnesium hydroxide particles (MH-Ca and MH-Ni samples) with a consequent decrease in sample porosity. In MH-Co case, the high density is due to a densely packed amorphous material formation. A correlation between morphological properties and the thermochemical behavior of $Mg(OH)_2$ is found. In particular, for all the investigated doped samples a lower reacted fraction is obtained in comparison with the not-doped material. However, because of the higher apparent density, the doped samples exhibit higher volumetric stored/released heat capacity. The highest value is reported for MH-Co3 sample (725 MJ/m^3), and it is almost three times higher than MH's value. In future development, the doped samples will be further investigated to enhance their performance, while maintaining the high density, and they will be tested for several cycles, to investigate their stability in real applications.

Author Contributions: Conceptualization, E.P.; Funding acquisition, E.P.; Investigation, M.F. and E.M.; Project administration, E.P.; Supervision, E.P.; Validation, M.F.; Visualization, E.P., M.F. and E.M.; Writing–original draft, E.P.; Writing–review and editing, E.P., M.F. and E.M.

Funding: This research was funded by INSTM (Consorzio Interuniversitario Nazionale per la Scienza e Tecnologia dei Materiali) grant number INSTMME002 "Thermochemical materials for heat storage: development and characterization".

Acknowledgments: This study was conducted as part of the project "Thermochemical materials for heat storage: development and characterization" sponsored by INSTM (Consorzio Interuniversitario Nazionale per la Scienza e Tecnologia dei Materiali), within the IEA SHC Task 58 "Material and Component Development for Thermal Energy Storage".

Conflicts of Interest: The authors declare no conflicts of interest. The funders had no role in: the design of the study; in the collection, analyses, or interpretation of data; in the writing of the manuscript, and in the decision to publish the results.

References

1. Tescari, S.; Singh, A.; Agrafiotis, C.; de Oliveira, L.; Breuer, S.; Schlögl-Knothe, B.; Roeb, M.; Sattler, C. Experimental evaluation of a pilot-scale thermochemical storage system for a concentrated solar power plant. *Appl. Energy* **2017**, *189*, 66–75. [CrossRef]
2. Li, G. Sensible heat thermal storage energy and exergy performance evaluations. *Renew. Sustain. Energy Rev.* **2016**, *53*, 897–923. [CrossRef]
3. De Gracia, A.; Cabeza, L.F. Phase change materials and thermal energy storage for buildings. *Energy Build.* **2015**, *103*, 414–419. [CrossRef]
4. Aydin, D.; Casey, S.P.; Riffat, S. The latest advancements on thermochemical heat storage systems. *Renew. Sustain. Energy Rev.* **2015**, *41*, 356–367. [CrossRef]
5. Zhang, H.; Baeyens, J.; Cáceres, G.; Degrève, J.; Lv, Y. Thermal energy storage: Recent developments and practical aspects. *Prog. Energy Combust. Sci.* **2016**, *53*, 1–40. [CrossRef]
6. Wu, J.; Long, X.F. Research progress of solar thermochemical energy storage. *Int. J. Energy Res.* **2015**, *39*, 869–888. [CrossRef]
7. Sarbu, I.; Sebarchievici, C. A Comprehensive review of thermal energy storage. *Sustainability* **2018**, *10*. [CrossRef]
8. Bowery, R.G.; Justen, J. Energy storage using the reversible oxidation of barium oxide. *Sol. Energy* **1978**, *21*, 523–525. [CrossRef]
9. Carrillo, A.J.; Sastre, D.; Serrano, D.P.; Pizarro, P.; Coronado, J.M. Revisiting the BaO_2/BaO redox cycle for solar thermochemical energy storage. *Phys. Chem. Chem. Phys.* **2016**, *18*, 8039–8048. [CrossRef] [PubMed]
10. Carrillo, A.J.; Serrano, D.P.; Pizarro, P.; Coronado, J.M. Manganese oxide-based thermochemical energy storage: Modulating temperatures of redox cycles by Fe–Cu co-doping. *J. Energy Storage* **2016**, *5*, 169–176. [CrossRef]
11. Carrillo, A.J.; Moya, J.; Bayón, A.; Jana, P.; de la Peña O'Shea, V.A.; Romero, M.; Gonzalez-Aguilar, J.; Serrano, D.P.; Pizarro, P.; Coronado, J.M. Thermochemical energy storage at high temperature via redox

cycles of Mn and Co oxides: Pure oxides versus mixed ones. *Sol. Energy Mater. Sol. Cells* **2014**, *123*, 47–57. [CrossRef]
12. Block, T.; Schmücker, M. Metal oxides for thermochemical energy storage: A comparison of several metal oxide systems. *Sol. Energy* **2016**, *126*, 195–207. [CrossRef]
13. Schmidt, M.; Szczukowski, C.; Roßkopf, C.; Linder, M.; Wörner, A. Experimental results of a 10 kW high temperature thermochemical storage reactor based on calcium hydroxide. *Appl. Therm. Eng.* **2014**, *62*, 553–559. [CrossRef]
14. Criado, Y.A.; Alonso, M.; Abanades, J.C. Conceptual process design of a CaO/Ca(OH)$_2$ thermochemical energy storage system using fluidized bed reactors. *Appl. Therm. Eng.* **2014**, *73*, 1089–1094. [CrossRef]
15. Criado, Y.A.; Alonso, M.; Abanades, J.C. Kinetics of the CaO/Ca(OH)$_2$ hydration/dehydration reaction for thermochemical energy storage applications. *Ind. Eng. Chem. Res.* **2014**, *53*, 12594–12601. [CrossRef]
16. Sakellariou, K.G.; Karagiannakis, G.; Criado, Y.A.; Konstandopoulos, A.G. Calcium oxide based materials for thermochemical heat storage in concentrated solar power plants. *Sol. Energy* **2015**, *122*, 215–230. [CrossRef]
17. Mastronardo, E.; Bonaccorsi, L.; Kato, Y.; Piperopoulos, E.; Lanza, M.; Milone, C. Thermochemical performance of carbon nanotubes based hybrid materials for MgO/H$_2$O/Mg(OH)$_2$ chemical heat pumps. *Appl. Energy* **2016**, *181*, 232–243. [CrossRef]
18. Shkatulov, A.; Aristov, Y. Thermochemical Energy Storage by LiNO$_3$-doped Mg(OH)$_2$: Dehydration Study. *Energy Technol.* Available online: https://onlinelibrary.wiley.com/doi/abs/10.1002/ente.201800050 (accessed on 14 May 2018).
19. Ryu, J.; Hirao, N.; Takahashi, R.; Kato, Y. Dehydration behavior of metal-salt-added magnesium hydroxide as chemical heat storage media. *Chem. Lett.* **2008**, *37*, 1140–1141. [CrossRef]
20. Müller, D.; Knoll, C.; Ruh, T.; Artner, W.; Welch, J.M.; Peterlik, H.; Eitenberger, E.; Friedbacher, J.; Harasek, M.; Blaha, P.; et al. Thermochemical Energy Storage: Calcium Doping Facilitates Water Dissociation in Magnesium Oxide. *Adv. Sustain. Syst.* **2018**, *2*, 1700096. [CrossRef]
21. Zamengo, M.; Junichi, R.Y.U.; Kato, Y. Chemical Heat Storage of Thermal Energy from a Nuclear Reactor by Using a Magnesium Hydroxide/Expanded Graphite Composite Material. *Energy Procedia* **2015**, *71*, 293–305. [CrossRef]
22. Piperopoulos, E.; Mastronardo, E.; Fazio, M.; Lanza, M.; Galvagno, S.; Milone, C. Enhancing the volumetric heat storage capacity of Mg(OH)$_2$ by the addition of a cationic surfactant during its synthesis. *Appl. Energy* **2018**, *215*, 512–522. [CrossRef]
23. Mastronardo, E.; Bonaccorsi, L.; Kato, Y.; Piperopoulos, E.; Lanza, M.; Milone, C. Strategies for the enhancement of heat storage materials performances for MgO/H$_2$O/Mg(OH)$_2$ thermochemical storage system. *Appl. Therm. Eng.* **2017**, *120*, 626–634. [CrossRef]
24. Mastronardo, E.; Kato, Y.; Bonaccorsi, L.; Piperopoulos, E.; Milone, C. Thermochemical Storage of Middle Temperature Wasted Heat by Functionalized C/Mg(OH)$_2$ Hybrid Materials. *Energies* **2017**, *10*, 70. [CrossRef]
25. Mastronardo, E.; Bonaccorsi, L.; Kato, Y.; Piperopoulos, E.; Milone, C. Efficiency improvement of heat storage materials for MgO/H$_2$O/Mg(OH)$_2$ chemical heat pumps. *Appl Energy* **2016**, *162*, 31–39. [CrossRef]
26. Inoue, M.; Hirasawa, I. The relationship between crystal morphology and XRD peak intensity on CaSO$_4$·2H$_2$O. *J. Cryst. Growth* **2013**, *380*, 169–175. [CrossRef]
27. Lv, Y.; Zhang, Z.; Lai, Y.; Li, J.; Liu, Y. Formation mechanism for planes (011) and (001) oriented Mg(OH)$_2$ films electrodeposited on SnO$_2$ coating glass. *Cryst. Eng. Comm.* **2011**, *13*, 3848–3851. [CrossRef]
28. Chen, D.; Zhu, L.; Zhang, H.; Xu, K.; Chen, M. Magnesium hydroxide nanoparticles with controlled morphologies via wet coprecipitation. *Mater. Chem. Phys.* **2008**, *109*, 224–229. [CrossRef]
29. Wu, Q.L.; Xiang, L.; Jin, Y. Influence of CaCl$_2$ on the hydrothermal modification of Mg(OH)$_2$. *Powder Technol.* **2006**, *165*, 100–104. [CrossRef]
30. Cordero, Z.C.; Schuh, C.A. Phase strength effects on chemical mixing in extensively deformed alloys. *Acta Mater.* **2015**, *82*, 123–136. [CrossRef]
31. Vanpoucke, D.E.P.; Cottenier, S.; Van Speybroeck, V.; Van Driessche, I.; Bultinck, P. Tetravalent Doping of CeO$_2$: The Impact of Valence Electron Character on Group IV Dopant Influence. *J. Am. Ceram. Soc.* **2014**, *97*, 258–266. [CrossRef]
32. Wu, J.S.; Du, J.; Gao, Y.M. Crystal growth morphology of magnesium hydroxide. *Turk. J. Chem.* **2014**, *38*, 402–412. [CrossRef]

33. Pardo, P.; Deydier, A.; Anxionnaz-Minvielle, Z.; Rougé, S.; Cabassud, M.; Cognet, P. A review on high temperature thermochemical heat energy storage. *Renew. Sustain. Energ. Rev.* **2014**, *32*, 591–610. [CrossRef]
34. Tang, X.; Guo, L.; Chen, C.; Liu, Q.; Li, T.; Zhu, Y. The analysis of magnesium oxide hydration in three-phase reaction system. *J. Solid State. Chem.* **2014**, *2013*, 32–37. [CrossRef]
35. Rocha, S.D.F.; Mansur, M.B.; Ciminelli, V.S.T. Kinetics and mechanistic analysis of caustic magnesia hydration. *J. Chem. Technol. Biotechnol.* **2004**, *79*, 816–821. [CrossRef]
36. Fruhwirth, O.; Herzog, G.W.; Hollerer, I.; Rachetti, A. Dissolution and hydration kinetics of MgO. *Surf. Technol.* **1985**, *24*, 301–317. [CrossRef]
37. Smithson, G.L.; Bakhshi, N.N. The kinetics and mechanism of the hydration of magnesium oxide in a batch reactor. *Can. J. Chem. Eng.* **1969**, *47*, 508–513. [CrossRef]

© 2018 by the authors. Licensee MDPI, Basel, Switzerland. This article is an open access article distributed under the terms and conditions of the Creative Commons Attribution (CC BY) license (http://creativecommons.org/licenses/by/4.0/).

Article

Solid-State Reactions for the Storage of Thermal Energy

Stefania Doppiu [1,*], Jean-Luc Dauvergne [1] and Elena Palomo del Barrio [1,2]

[1] Centro de Investigación Cooperativa de Energías Alternativas, CIC energiGUNE, 01510 Vitoria-Gasteiz, Spain; jldauvergne@cicenergigune.com (J.-L.D.); epalomo@cicenergigune.com (E.P.d.B.)
[2] Ikerbasque, Basque Foundation for Science, 48013 Bilbao, Spain
* Correspondence: sdoppiu@cicenergigune.com; Tel.: +34-945-297-108

Received: 20 December 2018; Accepted: 4 February 2019; Published: 7 February 2019

Abstract: In this paper, the use of solid-state reactions for the storing of thermal energy at high temperature is proposed. The candidate reactions are eutectoid- and peritectoid-type transitions where all the components (reactants and reaction products) are in the solid state. To the best of our knowledge, these classes of reactions have not been considered so far for application in thermal energy storage. This study includes the theoretical investigation, based on the Calphad method, of binary metals and salts systems that allowed to determine the thermodynamic properties of interest such as the enthalpy, the free energy, the temperature of transition, the volume expansion and the heat capacity, giving guidelines for the selection of the most promising materials in view of their use for thermal energy storage applications. The theoretical investigation carried out allowed the selection of several promising candidates, in a wide range of temperatures (300–800 °C). Moreover, the preliminary experimental study and results of the binary Mn-Ni metallic system are reported. This system showed a complex reacting behavior with several discrepancies between the theoretical phase diagram and the experimental results regarding the type of reaction, the transition temperatures and enthalpies and the final products. The discrepancies observed could be due both to the synthesis method applied and to the high sensitivity of the material leading to partial or total oxidation upon heating even if in presence of small amount of oxygen (at the ppm level).

Keywords: solid state reactions; thermal energy storage; nanocrystalline materials; ball milling

1. Introduction

The scope of this investigation is the development of performance materials with high energy density, reversibility, long cycle life, compact, low cost and with the potential to build "simple" thermal energy storage systems. This field of research is one of the priorities for i) helping the penetration and dispatchability of renewable energies; ii) contributing to create a low-carbon society for environmental protection; and iii) increasing energy efficiency and decreasing energy demand as targeted in the main road maps [1–3] related to energy policy and environment developed in recent years. In this context, the development of TES materials will play a major role, for example, in helping the re-utilization of wasted heat (e.g., in industrial processes) and in guaranteeing non-stop energy production, e.g., in solar energy power plants [4,5].

As is already well known, thermal energy can be stored using different processes: sensible, latent and thermochemical storage [5–9]. The energy capacity in these processes increases progressively from sensible to thermochemical processes. Unfortunately, this is accompanied by an increase of the complexity of the TES system that has to be developed, implying a substantial increase of the costs. For example, in the case of gas–solid reactions, the TES system should be composed of two reactors to keep the reactants separated (gas and solid) up to when the discharging process is needed (putting in contact the reactants to promote the exothermic reaction to recover the energy). This solution is technologically much more complex than the case of sensible storage, where the storage material, solid or liquid, is placed in a unique reactor. As a consequence, thermochemical storage is still at the prototyping/demonstration level and implies, so far, high investment costs. Latent heat storage (using solid–liquid or solid–solid phase transitions) is a more accessible technology nowadays at the demonstration level. Sensible storage is a mature technology already commercialized in many applications.

The main idea of this study is the use of solid-state chemical reactions as materials for thermal energy storage at high temperature. In particular, the focus is given to euteuctoid and peritectoid reactions [10] occurring in binary metals and salt-based systems. The goal was the identification of reactions fulfilling the requirements needed to be used as TES materials (high storage capacity, good thermal conductivity, mechanical and chemical stability, complete reversibility in charging/discharging cycles, affordable cost, etc.) and to obtain the experimental proof of their feasibility and reversibility.

These types of reactions have not been considered so far for the application addressed in this paper. The great advantages and novelties that they are potentially expected to bring in the TES field are as follows:

- Reactants and reaction products in the solid state can make it possible to conceive a TES system with the characteristic of simplicity the sensible storage system (only one reactor).
- Possible direct contact of the reacting material with the heat transfer fluid (allowing minimizing the use of expensive heat exchanger).
- Potential high energy density materials, allowing compact and low-cost systems.

These reactions offer many advantages, but can also present drawbacks, such as problems connected to the atomic diffusion in the solid-state (slow reaction kinetics) and the poor heat transfer rate in the solid-state especially when salt mixtures are taken into account.

The work included the selection of the most promising materials by a deep analysis of the existing databases of binary metals and salt systems [11], determining all the theoretical thermodynamic properties needed for the evaluation of the TES performances. As a result of the selection process, the Mn-Ni system was chosen for experimental investigation and feasibility study.

2. Materials and Methods

2.1. Materials Selection

The search for eutectoid and peritectoid reactions with suitable reaction temperatures was based on available phase diagrams for multi-component systems (ASM International, Scientific Group Thermodata Europe, ThemoTech Inc, etc. (GU2 7YG, Guildford, United Kingdom) and focused on metallic and salt binary systems. The theoretical performances were evaluated by using the CALPHAD (CALculation of PHAse Diagram) method. Regular and sub-regular solution models were used to obtain the Gibbs energy functions of various solution phases. The excess Gibbs energy of each phase was represented by the Redlich-Kister formalism, with binary interaction parameters following the form of power series [12–14]. These parameters were optimized by using the optimization module of FactSage7.0 software (7.0, GTT-Technologies, Herzogenrath, Germany) [15]. In particular, the selection was performed by using the set of evaluated and optimized thermodynamic databases for inorganic systems, such as light metal, alloy, molten salt, oxide. All the key thermodynamic properties (e.g., enthalpy of reaction, specific heats, densities, volume change during the charge/discharge process) of identified eutectoids and peritectoids were obtained assuming equilibrium conditions.

2.2. Nanocrystalline Materials Production

Two compositions in the Mn-Ni phase diagram were selected for experimental study: the peritectoid Mn_{75}-Ni_{25} and the eutectoid Mn_{52}-Ni_{48} (molar ratio). Mn and Ni powder were supplied by Alfa Aesar with purities of 99.3% and 99.8%, respectively. To avoid air contamination, the handling and sampling were carried out under controlled atmosphere in an Argon glove box (Brown) with levels of oxygen and humidity lower than 0.1 ppm.

To maximize the reactivity in the solid state, the materials were subjected to mechanochemical treatment (Ball milling) to achieve powders with a controlled degree of nanocrystallization. The mechanical milling was used only for the synthesis of nanostructured powder. The goal here was not to promote the reaction by ball milling but the preparation of highly reactive materials (high amount of defects, high specific surface area, and high contact area) and activate the reaction subsequently by thermal treatment.

For this purpose, a Spex mixer mill (875 RPM), using stainless steel vials and balls, was used. Two different milling procedures for the preparation of nanocrystalline materials were applied:

(1) Two-step synthesis: ball milling of separated Mn and Ni using 2 balls of 8 mm diameter. A milling time of 4 h was applied to the pure materials with a ball-to-powder mass ratio (BPR) of 1.6. To obtain a good intermixing between Mn and Ni, the materials were placed in the ball milling reactor in the right molar ratio (Mn_{75}-Ni_{25} and Mn_{52}-Ni_{48}) and subjected to mechanical treatment under mild conditions (for 15 min using 3 balls of 3 mm diameter).
(2) One-step synthesis: ball milling of Mn and Ni directly in the right molar ratio (75/25, 52/48) with 2 balls of 8 mm diameter. A milling time of 4 h was applied with a BPR of 1.6.

2.3. Structural Analysis

The structural analysis of the materials was performed by X-Ray diffraction analysis using a Bruker D8 Discover equipped with a LYNXEYE XE detector with monochromatic Cu Kα1 radiation of λ = 1.54056 Å. Patterns were recorded in a 2θ angular range 10–120° with a step size of 0.02° and a step time of 1.5 s. The measurements were performed at room temperature. The structural evolution upon heating was studied by in situ XRD measurements by using a Bruker Advance D8 instrument with cobalt radiation ($\lambda Co\alpha 1$ = 1.78886 Å/$\lambda Co\alpha 2$ = 1.79277 Å). The equipment operated in Brag-Brentano theta-theta geometry, with an operating power of 30 kV and 50 mA. The samples were placed in a nickel-coated high-vacuum chamber designed for the use in the range from room temperature up to 1200 °C (HTK 1200N) under a high vacuum, inert and reactive atmosphere. The sample was mounted on an alumina sample holder avoiding any contact with the wall of the chamber and in contact with the temperature sensor.

Information about the phases formed and their relative percentages, the crystallite sizes and the microstrain level were obtained from the X-ray patterns by using a full profile fitting procedure [16] based on the Rietveld method [17].

The morphology of the material was studied by Scanning Electron Microscopy (SEM,) using a Quanta 200 FEG scanning electron microscope (FEI Company, Hillsboro, OR, USA) operated in high-vacuum mode at 30 kV and with a back-scattered electron detector (BSED). In addition, energy-dispersive X-ray spectroscopy (EDX) analyses were carried out in order to obtain chemical composition maps.

2.4. Reactivity and Thermodynamic Characterization

The reactivity of the materials was tested by Differential Scanning Calorimetry (DSC) technique using a Thermal Analysis Q2000 model. These techniques allowed the determination of the reaction temperatures and the reaction enthalpies. For all the measurements the heating rate was 5 K/min with three or twenty cycles between 450 and 660 °C including isothermal steps of 30 min between subsequent heating and cooling processes. The structural changes of the materials after DSC experiments were determined by XRD analysis.

3. Results and Discussion

3.1. Materials Selection Results

More than 200 binary phase diagrams (metals and salts) were analyzed using available databases and the FactSage7.0 software (7.0, GTT-Technologies, Herzogenrath, Germany) in the temperature range of 300–1000 °C. The criteria of selection were based on availability of the materials, no toxicity and relatively low cost. The modelling made it possible to identify all the transitions of interest in the systems studied (eutectoids, peritectoids, peritectics, eutectics, etc.), together with the associated energy densities and main thermodynamic and thermophysical parameters. In this study, all the systems with eutectoid and peritectoid transitions with theoretical volumetric energy densities lower than 100 kWh/m^3 were discarded. The results of the theoretical modelling allowed the selection of several potential candidates, in a wide range of temperatures, shown in Table 1, that could be used for further experimental investigation.

Table 1. Results of the selection process.

System	Temperature °C	ΔHr J/mol	Volumetric Energy Density [kWh/m3]	Volumetric Energy Density (ΔT = 100 K)[kWh/m3]
Mn/Ni (52/48)	251	4281	166	281
Cu/Sn (84/16)	505	1505	48	98
Mn/Zn (68/32)	529	1407	48	149
Mn/Ni (25/75)	532	3746	171	331
Mn/Ni (75/25)	566	16774	583	707
Li$_2$SO$_4$/Na$_2$SO$_4$ (50/50)	515	40080	217	319
Al/Ni (38/62)	698	1504	52	176
Cu/Sn (75/25)	676	3124	84	465 (ΔT = 58 K)
Fe/Sn (60/40)	612	4011	97.7	176
Mn/Ni (41/59)	607	2784	117	265
Al/Mn (45/55)	842	6708	218	342
Fe/Si (30/70)	960	4996	165	263
Fe/Si (67/33)	962	3824	121	226

Another aspect considered in this study was how to compare these reactions with other types of thermal energy storage processes (sensible, latent or thermochemical), thinking about their possible integration into a real application. To that end, several aspects have to be considered: (i) we deal with chemical reactions where all the components, reagents and products, are in the solid state; (ii) the reaction mechanism is governed by the atomic diffusion in the solid state; and (iii) the reaction occurs at a well-defined constant temperature. Considering the solid-state nature of these materials, they can be assimilated to a sensible storage material with the difference that at a certain temperature an extra contribution to the sensible heat is given by the enthalpy of the reaction. As a result, depending on the reaction and its energy and on the thermophysical properties, such as the Cp, for the reactions proposed in this paper a higher overall energy density is expected. In Figure 1, the energy density obtained by the sum of the sensible heat contribution and the reaction enthalpy, considering a range of temperature of 100 K around the reaction temperature, is compared to the best sensible storage materials.

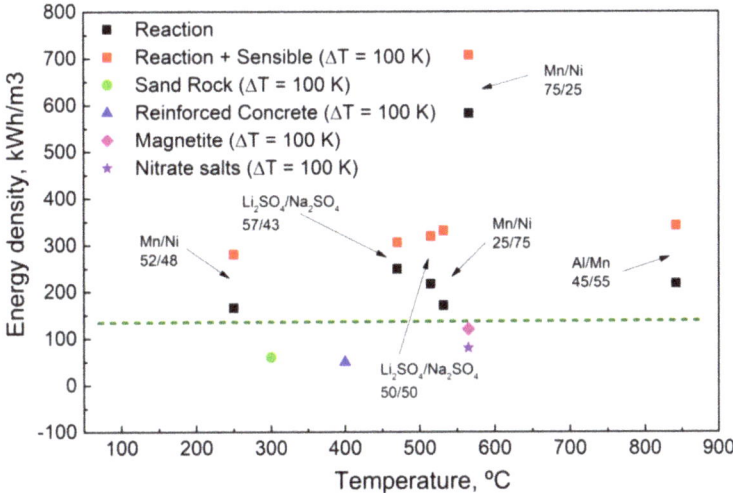

Figure 1. Theoretical volumetric energy densities of some selected solid-state reactions (black squares). The energy corresponding to a ΔT of 100 K (red squares), together with the values relevant to the best sensible storage materials (ΔT = 100 K), is also pictured.

This allows achieving, for almost all the systems considered, theoretical energy densities between 250–350 kWh/m^3. These values are higher than the sensible storage materials considered nowadays. For example, the magnetite, the best material identified so far, presents an energy density of 120 kWh/m^3 for a ΔT = 100 K. These results are very promising and confirm the great theoretical potential of these reactions. It is noteworthy that the Mn$_{75}$-Ni$_{25}$ shows theoretical energy densities considerably higher compared to the other systems (583 kWh/m^3 for the reaction and 707 kWh/m^3 when a ΔT of 100 K is considered). Due to these results, the Mn-Ni system was the first choice for the experimental investigation.

To determine univocally the composition with the highest energy density, for each system studied and each transition of interest, four compositions around the theoretical one were analyzed (two before and two after). In particular, the Mn-Ni system presents three solid-state reactions (both eutectoid and peritectoid) below 600 °C at the compositions Mn$_{75}$-Ni$_{25}$, Mn$_{52}$-Ni$_{48}$ and Mn$_{25}$-Ni$_{75}$ with promising theoretical energy densities as shown in Figure 2.

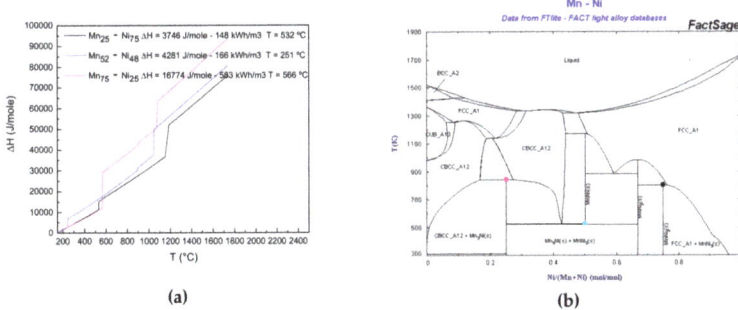

Figure 2. (a) Theoretical energy densities of the solid-state reactions corresponding to the compositions Mn$_{75}$-Ni$_{25}$, Mn$_{52}$-Ni$_{48}$ and Mn$_{25}$-Ni$_{75}$. The relative volumetric energy densities are also reported in the Figure. In (b), the Mn-Ni phase diagram is shown, with the corresponding transition highlighted (colored dots).

3.2. Synthesis of Mn$_{75}$-Ni$_{25}$ and Mn$_{52}$-Ni$_{48}$ Nanocrystalline Materials

The reactions studied in this paper are governed by the diffusion in the solid state. As a consequence, it is imperative to find synthesis routes in order to maximize the atomic diffusion by decreasing the atomic diffusion path length (small grain sizes as well as small particle sizes), introducing structural defects (dislocation, grain boundaries step, kink and corner atoms, etc.) and promoting high intermixing degree to guarantee the maximum contact between the reagents (high specific surface area). It is well known that a powerful tool for achieving these results is given by mechanochemical techniques [18]. The two procedures applied for the synthesis of nanocrystalline Mn$_{75}$-Ni$_{25}$ and Mn$_{52}$-Ni$_{48}$ led to the formation of samples with different microstructures and similar degree of nanocrystallization (grain sizes around 15 nm) and, as wanted, with no evidence of the formation of the intermetallic Mn$_3$Ni and MnNi predicted in the phase diagram. The two compositions studied showed different behavior depending on the milling treatment applied. In Figure 3, the XRD patterns for the four samples are reported together with the results of the fitting procedure using the Rietveld method. Only the Mn and Ni reflections are detected in the XRD patterns.

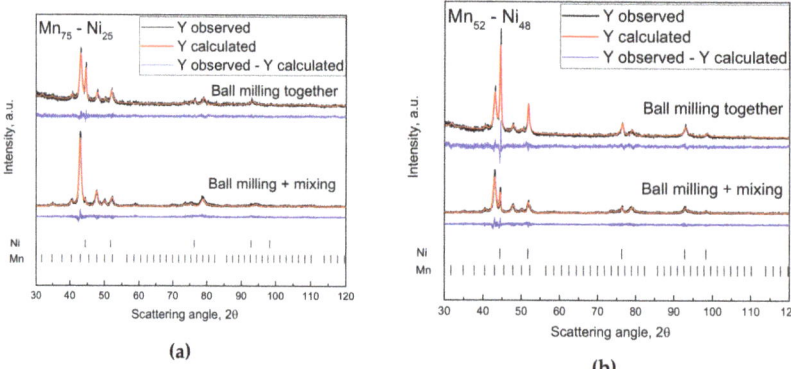

Figure 3. X-ray diffraction results of the samples (**a**) Mn$_{75}$-Ni$_{25}$ and (**b**) Mn$_{52}$-Ni$_{48}$ prepared by one-step or two-step synthesis. The results of the fitting procedure using the Rietveld method are also reported. Black line: experimental pattern. Red line: theoretical pattern. Blue line: difference between experimental and theoretical patterns.

The two-step synthesis caused, for all Mn$_{75}$-Ni$_{25}$ samples tested in this investigation, the gradual disappearance of the diffraction peaks corresponding to Nickel (PDF number: 01-071-3740 4-850) (see Figure 3a pattern below). Surprisingly, the gradual disappearance of Ni reflections is not accompanied by a substantial variation of the cell parameters of Mn (PDF number: 01-089-2105 32-637), as a consequence of the substitution of Mn atoms by Ni in the Mn net with the formation of a Mn$_{1-x}$Ni$_x$ solid solution (see Table 1). The solubilization of Ni into Mn should lead to the decrease of the volume of the primitive cell due to the smaller size of Ni compared to Mn (reference value for Mn being the sample milled 4 h, see Table 2).

Table 2. Results of the Rietveld analysis for Mn$_{75}$-Ni$_{25}$ and Mn$_{52}$-Ni$_{48}$.

4 h Ball Milling	a (nm)	Error (nm)	<d> (nm)	Error (nm)	ε strain	Error
Mn (4 h)	0.89088	±4.1 × 10^{-5}	19.3	±0.20	3.7 × 10^{-3}	±9.2 × 10^{-5}
Mn (Mn$_{75}$-Ni$_{25}$ One-step synthesis)	0.88802	±7.7 × 10^{-5}	14.5	±0.27	1.8 × 10^{-3}	±2.5 × 10^{-4}
Mn (Mn$_{75}$-Ni$_{25}$ Two-step synthesis)	0.89107	±4.4 × 10^{-5}	16.6	±0.16	2.6 × 10^{-3}	±8.0 × 10^{-5}
Mn (Mn$_{52}$-Ni$_{48}$ One-step synthesis)	0.88879	±9.7 × 10^{-5}	14.4	±0.33	1.4 × 10^{-3}	±4.4 × 10^{-4}
Mn (Mn$_{52}$-Ni$_{48}$ Two-step synthesis)	0.89039	±9.7 × 10^{-6}	17.2	±0.22	3.7 × 10^{-3}	±6.6 × 10^{-5}

The sample prepared by one-step synthesis (where the Ni is clearly visible after 4 h of milling, Figure 3a pattern above) shows the expected behavior (decreasing of the Mn cell parameter), while the sample prepared by two-step synthesis (where the Ni is hardly visible after 4 h of milling, Figure 3a pattern below) shows an opposite trend, with slightly higher cell parameters compared to the Mn milled for 4 h. Further studies were then carried out, by applying mechanical milling progressively higher (30 min, 1 h, 2 h, 4 h and 8 h), in order to clarify this behavior and correlate the disappearing of Ni reflection with the structural modification of Mn. For higher milling times (8 h and 16 h) no Ni was detected in the mixture after mixing (XRD). In Figure 4, the SEM micrograph with the corresponding EDX analysis of the sample milled for 8 h are reported, confirming the results obtained by XRD investigation.

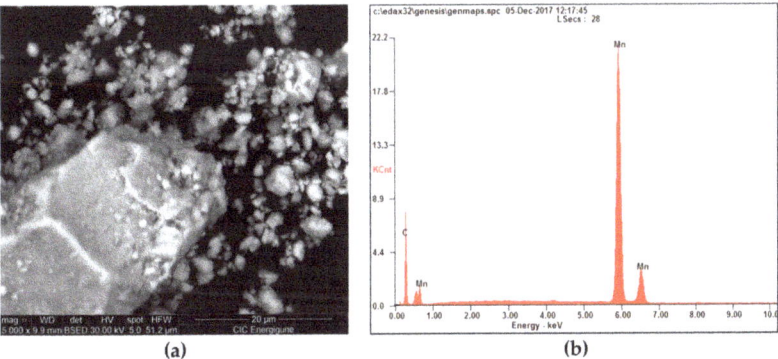

Figure 4. SEM micrograph (**a**) and EDX analysis (**b**) of the samples Mn_{75}-Ni_{25} (two-step synthesis).

For this sample, the particles size distribution was determined by following two approaches a) using a particles size analyzer (Master sizer 3000, Malvern), and b) using the software ImageJ (version 2.0, an open source Java-based software) [19,20]. Following the two approaches, similar results were obtained within media of around 4 μm (SEM/ImageJ) and around 6 μm (particle size analyzer). The difference between the crystallite sizes and the particle sizes is not surprising, due to the phenomena occurring during milling that, in the case of metals, can promote the cold welding of the particles with a consequent increase of their sizes without decreasing the overall reactivity of the material if the crystallite sizes remain small. The behavior under milling and the final structure obtained, depending on the conditions applied, are now subject to deep study due to the very different reacting behavior observed for the different samples, as will be explained later in the text.

3.3. Reactivity upon Heating

To study the reactivity upon heating and cooling, all the samples synthesized were studied by differential scanning calorimetry and simultaneous thermal analysis techniques. The goal was to determine the reactivity, to quantify the energy relative to the solid-state reaction (peritectoid, eutectoid), and to perform preliminary study on the reversibility by cycling test. The results of the DSC tests are shown in Figure 5.

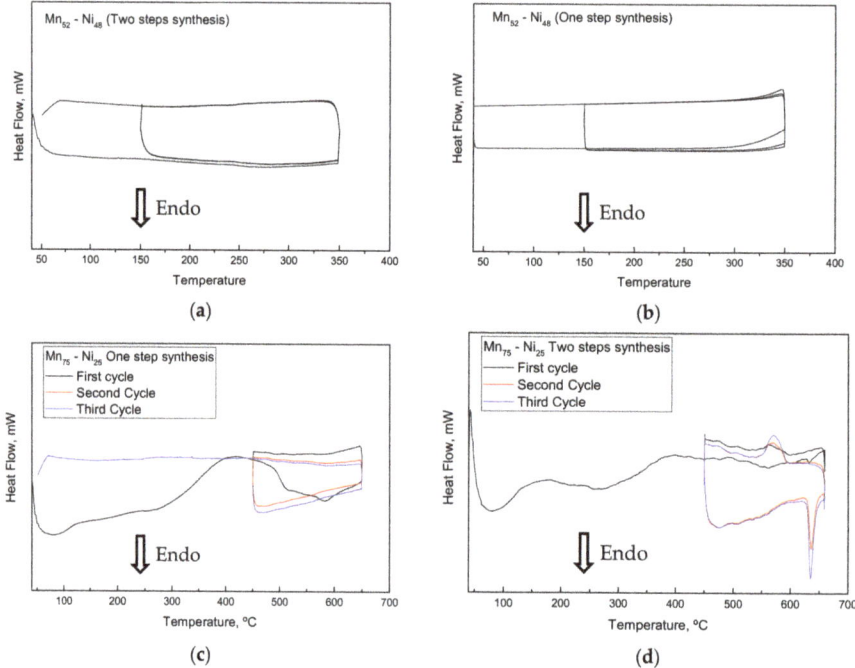

Figure 5. DSC results of the samples: (**a**) Mn_{52}-Ni_{48} prepared by two-step synthesis, (**b**) Mn_{52}-Ni_{48} prepared by one-step synthesis, (**c**) Mn_{75}-Ni_{25} prepared by two-step synthesis, and (**d**) Mn_{75}-Ni_{25} prepared by one-step synthesis.

Regarding the DSC results several aspects can be highlighted:

There was a discrepancy between the theoretical phase diagram (see inset Figure 1) and the experimental results for the composition Mn_{52}-Ni_{48}, where no reactivity was detected in that range of composition between room temperature and 350 °C (eutectoid reaction expected at 251 °C). This result was confirmed by further experiments carried out under the same experimental conditions.

The reacting behavior of the composition Mn_{75}-Ni_{25} is strongly influenced by the experimental conditions applied for the synthesis of the material (no reactivity in case of one-step synthesis).

Following the theoretical phase diagram at the composition Mn_{75}-Ni_{25}, the reaction between Mn and Ni should lead to the formation of the intermetallic Mn_3Ni. This reaction was expected to be exothermic, while an endothermal event was observed in the experimental results at higher temperature than the predicted one (630 °C instead of 566 °C). Moreover, the XRD patterns performed after DSC measurements reveal, for most of the samples, only the presence of Mn, $Mn_{(1-x)}Ni_xO$ and traces of Ni. The detection of the formation of a small amount of one new "unknown phase" was possible only after many experiments, where the extensive oxidation of the sample during the DSC experiment was limited.

The reaction is reversible with a progressive increase of the enthalpy upon cycling. It is noteworthy that the value of the enthalpy for this system cannot be given precisely due to the very high reactivity of the material upon heating, leading to oxidation even when controlled atmosphere (level of oxygen below 0.1 ppm) or vacuum is applied. The energy obtained experimentally is considerably lower than the predicted one (around 10 J/g instead of 300 J/g). This behavior is amplified and clearly visible by performing cycling experiments in the DSC apparatus (up to 20 cycles), as shown in Figure 6.

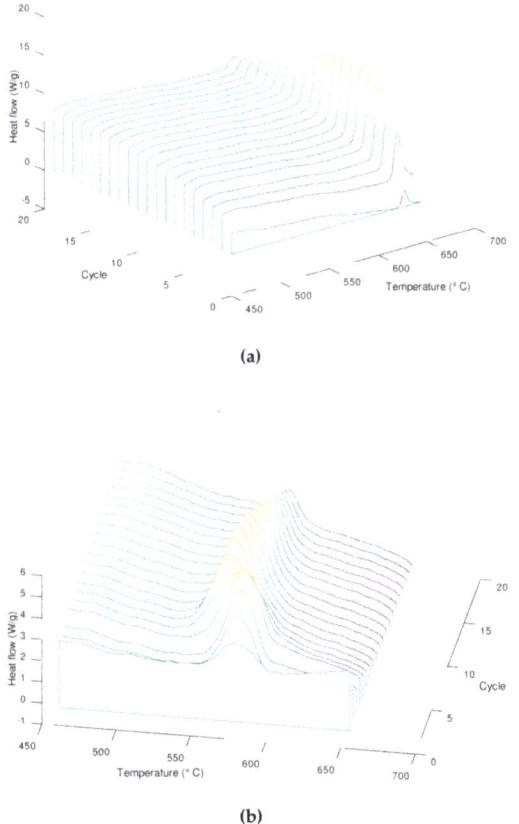

Figure 6. Cycling results of the sample Mn$_{75}$-Ni$_{25}$ synthesized by the two-step synthesis. (**a**) Heating steps, (**b**) cooling steps.

The cycling results show the progressive increase of the enthalpy of the reaction up to a certain limit, when the enthalpy starts to decrease to very low values (the DSC peak almost disappears). During the cooling process, a partial displacement of the peak during the first cycles can also be observed, reaching a stationary regime after eight cycles. In these experiments, two competitive effects are occurring at the same time, the progressive oxidation of the sample, which causes the increase of the inactive phase in the mixture (no contribution to the reaction heat), and the increase of the reaction enthalpy due to the intrinsic behavior of the mixture (still under investigation).

The X-ray diffraction analysis after the cycling experiment (20 cycles) did not allow correlation of the final structure with the reactivity observed. Only in the case of one of the samples cycled three times was it possible to obtain an XRD diffractogram (reported in Figure 7) in which the formation of a new unknown phase was detected (diffraction peaks at 2θ angular range of 36.2°, 47.2° and 68.9°). This phase could correspond to the intermetallic Mn$_3$Ni predicted in the phase diagram; unfortunately, no crystallographic information is available for this phase, hindering its univocal determination.

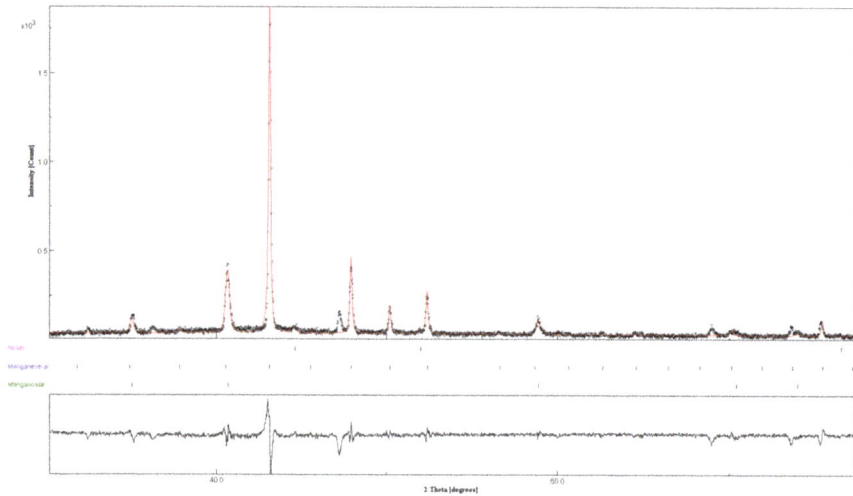

Figure 7. X-ray diffraction results of Mn_{75}-Ni_{25} sample (two-step synthesis) after DSC experiment.

To shed light on the reacting behavior observed and to determine the phenomena occurring upon heating several "in situ" XRD experiments were planned and carried out applying the same heating protocol used in the DSC experiments. The goal was to correlate the transitions observed in the DSC with the corresponding structural modification in the mixture. Unfortunately, none of the attempts made in order to obtain these measurements were successful, due to the extremely high reactivity of the mixture in the presence of traces of oxygen. Different trials were carried out under vacuum, under overpressure of N_2 and in dynamic atmosphere (continuous vacuum and N_2 flux). In all cases, the complete oxidation of the Mn_{75}-Ni_{25} mixture was observed with the formation of the mixed oxide $Mn_{1-x}Ni_xO$ (see Figure 8). It is noteworthy that, for the composition studied, the formation of the two solid solutions $Mn_{1-x}Ni_xO$ and $Ni_{1-x}Mn_xO$ should be detected, while only one phase was detected [21].

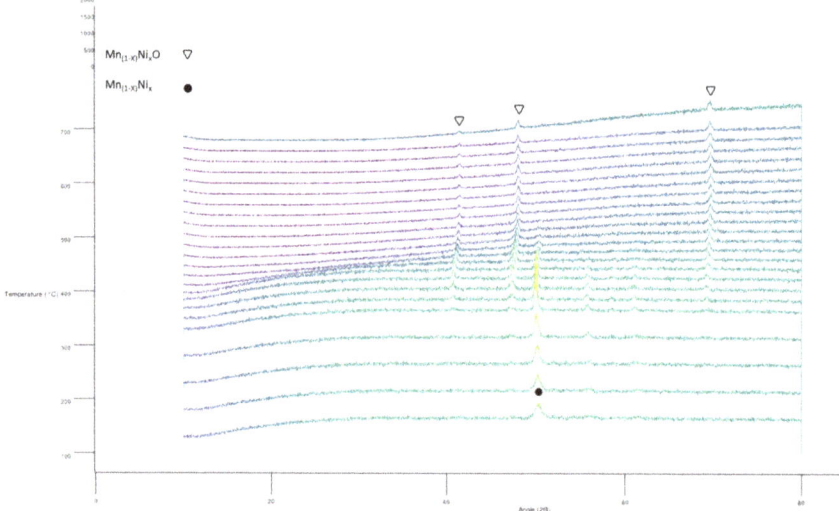

Figure 8. "In situ" X-ray diffraction measurements of Mn_{75}-Ni_{25} sample (two-step synthesis).

The same problems of oxidation were encountered when trying to perform measurements to test the thermophysical properties of Mn_{75}-Ni_{25} upon heating (for examples in the case of LFA measurements), not making it possible to obtain further experimental results.

4. Conclusions and Perspectives

In this paper, solid-state reactions are proposed as possible candidates for thermal energy storage applications at high temperature. This study allowed the selection of several reactions with theoretical energy density above 100 kWh/m^3. Two Mn-Ni compositions (Mn_{75}-Ni_{25} and Mn_{52}-Ni_{48}) were chosen for the experimental study. The reaction was activated by thermal treatment after the preliminary preparation of nanostructured powders by ball milling techniques. The behavior of this system revealed considerably more complex than expected. More and more questions arose with the proceeding of the experimental investigation. For example, i) it is well known that mechanical milling is a powerful technique to increase the solubility limit in binary metallic systems; however, the solubilization of considerably high amount of nickel in the manganese net was accompanied by only a slight variation of the cell parameters. ii) A high degree of solubilization of Ni into Mn in the case of one-step synthesis where the two elements were milled at high milling intensity for up to 4 h was expected, while the results show an opposite behavior, with a higher solubilization degree being reached in the mixture prepared by two-step synthesis (the two elements milled together only for 15 min in mild conditions). iii) The direct milling of the two elements (one-step synthesis) led to a total absence of reactivity, even if similar degrees of nanocrystallization and homogenization were achieved, compared to the two-step synthesis. In addition, finally, what type of reacting event is connected to the reversible and progressively increasing peak observed in the DSC analysis? We are now pursuing different strategies to understand all of the behavior observed. The work is more focused on the material science point of view to explain the peculiar reactivity observed. Regarding the performance of this material for thermal energy storage applications, two main aspects can be considered. On one hand, the discrepancies between the theoretical reaction enthalpy and the experimental one are probably due to the progressive oxidation of the sample during thermal cycling. In this respect, it is hard to be definitive with regard to the performance as thermal energy storage material up to when it will be possible to test the material avoiding its oxidation. On the other hand, a material so reactive is not suitable for thermal energy storage applications because of the extremely controlled experimental conditions needed to avoid its degradation. This could be an important technological constraint leading, most probably, to investment costs that are too high for large-scale applications.

Author Contributions: Conceptualization, S.D. E.P.; Methodology, S.D.; Software, J-L.D.; Validation, S.D., E.P.; Investigation, S.D.; Resources, E.P.; Data curation, S.D., J-L.D., E.P.; Writing—original draft preparation, S.D.; Writing—review and editing, S. D., J-L.D., E.P.; Visualization, S.D., J-L.D.; Project administration, S.D.; Funding acquisition, S.D., E.P.

Funding: This research was funded by the European Union's Horizon 2020 research and innovation programme under the Marie Skłodowska-Curie grant agreement No 752520". SOLSTORE project (Solid-state reactions for thermal energy storage).

Acknowledgments: The authors acknowledge Cristina Luengo Vilumbrales and Maria Jáuregui for the help and commitment in the experimental measurements.

Conflicts of Interest: The authors declare and confirm that there is no conflict of interest.

References

1. Materials Roadmap Enabling Low Carbon Energy Technologies. Available online: http://ec.europa.eu/research/industrial_technologies/pdf/materials-roadmap-elcet13122011_en.pdf (accessed on 19 December 2018).
2. Energy Roadmap 2050. Available online: http://eur-lex.europa.eu/LexUriServ/LexUriServ.do?uri=COM:2011:0885:FIN:EN:PDF (accessed on 19 December 2018).

3. EASE/EERA Energy Storage Technology Development Roadmap towards 2030. Available online: http://ease-storage.eu/easeeera-energy-storage-technology-development-roadmap-towards-2030 (accessed on 19 December 2018).
4. BCS Incorporated (March 2008). Available online: http://www1.eere.energy.gov/manufacturing/intensiveprocesses/pdfs/waste_heat_recovery.pdf (accessed on 19 December 2018).
5. Siegel, N.P. Thermal energy storage for solar power production. *Energy Environ.* **2012**, *1*, 119–131. [CrossRef]
6. Lane, G.A.; Shamsundar, N. *Solar heat storage: Latent heat materials, Vol. I: Background and scientific principles.*; American Society of Mechanical Engineers: New York, NY, USA, 1983; p. 467. [CrossRef]
7. Zalba, B.; Marın, J.M.; Cabeza, L.F.; Mehling, H. Review on thermal energy storage with phase change: materials, heat transfer analysis and applications. *Appl. Therm. Eng.* **2003**, *23*, 251–283. [CrossRef]
8. Felderhoff, M.; Urbanczyk, R.; Peil, S. Thermochemical heat storage for high temperature applications—A review. *Green* **2013**, *3*, 113–123. [CrossRef]
9. Steinfeld, A.; Palumbo, R. Solar thermochemical process technology. *Encycl. Phys. Sci. Technol.* **2001**, *15*, 237–56. [CrossRef]
10. Cahn, R.W.; Haasen, P. *Physical metallurgy: fourth, revised and enhanced edition*; Elsevier Science BV: Amsterdam, The Netherlands, 1996.
11. Massalski, T.B. *Binary alloy phase diagrams*; ASM international: Materials Park, OH, USA, 1992.
12. Saunders, N.; Miodownik, A.P. *CALPHAD (calculation of phase diagrams): a comprehensive guide*; Elsevier: New York, NY, USA, 1998.
13. Chang, Y.A.; Chen, S.; Zhang, F.; Yan, X.; Xie, F.; Schmid-Fetzer, R.; Oates, W.A. Phase diagram calculation: past, present and future. *Prog. Mater Sci.* **2004**, *49*, 313–345. [CrossRef]
14. Cacciamani, G. An Introduction to the CALPHAD Method and the COMPOUND Energy Formalism (CEF). *Tecnol. Metal. Mater. Mineração* **2016**, *13*, 16. [CrossRef]
15. Bale, C.W.; Bélisle, E.; Chartrand, P.; Decterov, S.A.; Eriksson, G.; Gheribi, A.E.; Hack, K.; Jung, I.H.; Kang, Y.B.; Melançon, J.; et al. FactSage Thermochemical Software and Databases-2010-2016. *Calphad* **2016**, *54*, 35–53. [CrossRef]
16. Lutterotti, L.; Gialanella, S. X-ray diffraction characterization of heavily deformed metallic specimens. *Acta Mater.* **1998**, *46*, 101–110. [CrossRef]
17. Rietveld, H. A profile refinement method for nuclear and magnetic structures. *J. Appl. Crystallogr.* **1969**, *2*, 65–71. [CrossRef]
18. Suryanarayana, C. Structure and properties of nanocrystalline materials. *Bull. Mater. Sci.* **1994**, *17*, 307–346. [CrossRef]
19. Schneider, C.A.; Rasband, W.S.; Eliceiri, K.W. NIH Image to ImageJ: 25 years of image analysis. *Nat. Methods* **2012**, *9*, 671–675. [CrossRef] [PubMed]
20. Igathinathane, C.; Pordesimo, L.O.; Columbus, E.P.; Batchelor, W.D.; Methuku, S.R. Shape identification and particles size distribution from basic shape parameters using ImageJ. *Comput. Electron. Agric.* **2008**, *63*, 168–182. [CrossRef]
21. Bergman, B.; ÅGREN, J. Thermodynamic Assessment of the System MnO-NiO. *J. Am. Ceram. Soc.* **1985**, *68*, 444–450. [CrossRef]

© 2019 by the authors. Licensee MDPI, Basel, Switzerland. This article is an open access article distributed under the terms and conditions of the Creative Commons Attribution (CC BY) license (http://creativecommons.org/licenses/by/4.0/).

MDPI
St. Alban-Anlage 66
4052 Basel
Switzerland
Tel. +41 61 683 77 34
Fax +41 61 302 89 18
www.mdpi.com

Nanomaterials Editorial Office
E-mail: nanomaterials@mdpi.com
www.mdpi.com/journal/nanomaterials

www.ingramcontent.com/pod-product-compliance
Lightning Source LLC
LaVergne TN
LVHW070543100526
838202LV00012B/364